Physical Science

> ## THIS PRODUCT IS INTENDED FOR HOME USE ONLY
>
> The images and all other content in this book are copyrighted material owned by Elemental Science, Inc. Please do not reproduce this content on email lists or websites. If you have an eBook, you may print out as many copies as you need for use WITHIN YOUR IMMEDIATE FAMILY ONLY. Duplicating this book or printing the eBook so that the book can then be reused or resold is a violation of copyright.
>
> **Schools and co-ops:** You MAY NOT DUPLICATE OR PRINT any portion of this book for use in the classroom. Please contact us for licensing options at support@elementalscience.com.

Physical Science

First Edition, 2021

Copyright @ Elemental Science, Inc.

Email: support@elementalscience.com

ISBN 978-1-953490-07-0

Printed in the USA for worldwide distribution

For more copies write to:
Elemental Science
PO Box 79
Niceville, FL 32588
support@elementalscience.com

Copyright Policy

All contents copyright © 2021 by Elemental Science. All rights reserved.

Limit of Liability and Disclaimer of Warranty: The publisher has used its best efforts in preparing this book, and the information provided herein is provided "as is." Elemental Science makes no representation or warranties with respect to the accuracy or completeness of the contents of this book and specifically disclaims any implied warranties of merchantability or fitness for any particular purpose and shall in no event be liable for any loss of profit or any other commercial damage, including but not limited to special, incidental, consequential, or other damages.

Trademarks: This book identifies product names and services known to be trademarks, registered trademarks, or service marks of their respective holders. They are used throughout this book in an editorial fashion only. In addition, terms suspected of being trademarks, registered trademarks, or service marks have been appropriately capitalized, although Elemental Science cannot attest to the accuracy of this information. Use of a term in this book should not be regarded as affecting the validity of any trademark, registered trademark, or service mark. Elemental Science is not associated with any product or vendor mentioned in this book.

This Guide at a Glance

Dear Student,

Welcome to your Physical Science course! This program is written to guide you, the student, through this course. Here is a quick look at what you will find inside.

1. Textbook Assignments

Know what to read in the free, digital textbook, *CK-12 Physical Science*. Download the version used in the guide:

🖥 https://elementalscience.com/blogs/resources/ps

2. Experiment Information

If you choose the hands-on option, your weekly experiment directions will be found in this section. If you choose the digital option, the online lab information directly follows the experiment.

3. Events in Science

Learn about current events in science and build your internet research skills with these assignments.

4. Hands-on Activity

Have some extra science fun with these optional weekly hands-on STEAM activities.

5. & 6. Course Options

Choose from a hands-on option (5) or a digital option (6) for completing this course. Either way, these scheduling grids will make planning your weekly science coursework a snap! These schedule sheets lay out your textbook, experiment, and writing assignments for the week. Plus, each one shares the main objective for the week along with any supplies you may need.

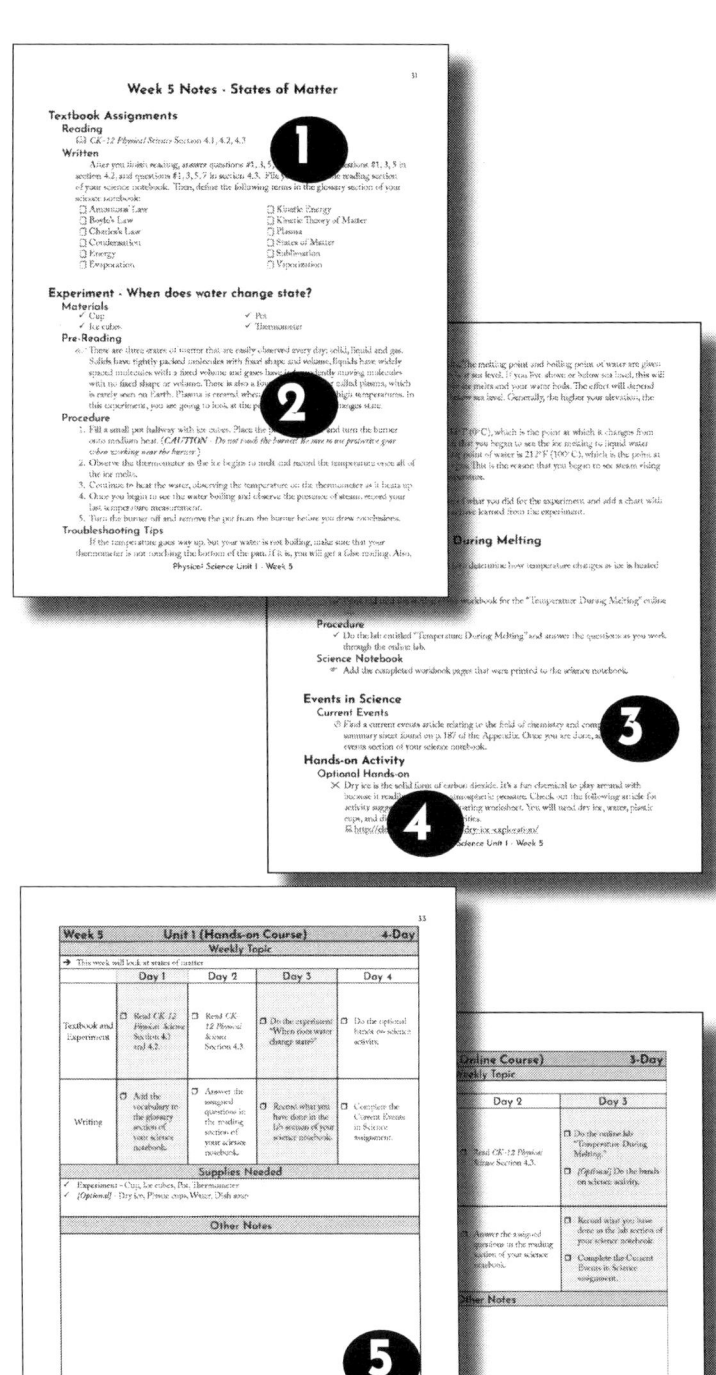

Physical Science Table of Contents

Introduction

Introduction for the Student	7
Adding Current Events into Your Science Studies	9
A Note to the Facilitator	10

Unit 1: Matter

Unit 1 – Overview of Study	13
Week 1 Notes – What Is Science?	14
Week 2 Notes – Scientific Research, Part 1	18
Week 3 Notes – Scientific Research, Part 2	22
Week 4 Notes – Matter	26
Week 5 Notes – States of Matter	31
Week 6 Notes – Atoms	35
Week 7 Notes – The Periodic Table	39
Unit 1 – Answers	43

Unit 2: Chemical Properties

Unit 2 – Overview of Study	47
Week 1 Notes – Chemical Bonding, Part 1	48
Week 2 Notes – Chemical Bonding, Part 2	52
Week 3 Notes – Chemical Reactions, Part 1	56
Week 4 Notes – Chemical Reactions, Part 2	60
Week 5 Notes – Chemistry of Carbon, Part 1	64
Week 6 Notes – Chemistry of Carbon, Part 2	68
Week 7 Notes – Chemistry of Solutions	73
Week 8 Notes – Nuclear Chemistry	78
Unit 2 – Answers	81

Unit 3: Forces and Motion

Unit 3 – Overview of Study	85
Week 1 Notes – Motion	86
Week 2 Notes – Forces, Part 1	91
Week 3 Notes – Forces, Part 2	96
Week 4 Notes – Newton's Law of Motion	100
Week 5 Notes – Fluid Forces	105
Week 6 Notes – Work and Machines, Part 1	109
Week 7 Notes – Work and Machines, Part 2	113
Unit 3 – Answers	117

Unit 4: Energy and Electricity

Unit 4 – Overview of Study	121
Week 1 Notes – Introduction to Energy	122
Week 2 Notes – Thermal Energy	127
Week 3 Notes – Waves	132
Week 4 Notes – Sound	136
Week 5 Notes – Electromagnetic Radiation	140
Week 6 Notes – Visible Light	144
Week 7 Notes – Electricity, Part 1	148
Week 8 Notes – Electricity, Part 2	152
Week 9 Notes – Magnetism	156
Week 10 Notes – Electromagnetism	160
Unit 4 – Answers	164

Unit 5: The Science Fair Project

Unit 5 – Overview of Study	167
Week 1 Notes – Steps 1 & 2	168
Week 2 Notes – Steps 3 & 4	172
Week 3 Notes – Steps 5 & 6	176
Week 4 Notes – Steps 7 & 8	180

Appendix

Scientist Biography Report Grading Rubric	185
Science in the News Template	187

Density Calculations	188
Element Profile Page	190
Nomenclature Worksheet	192
Nomenclature Practice	193
Chemical Equations Worksheet	194
Resultant Force Worksheet	197
Second Law of Motion Worksheet	198
Power Worksheet	199
Mechanical Wave Worksheet	200
Electromagnetic Wave Worksheet	201
Electrical Charge Worksheet	202

Introduction for the Student

Welcome to *Physical Science*! This year, you will learn about matter, chemical reactions, motion, forces, thermal energy, and much more. In this guide, you will find two types of schedules and notes with the assignments for each week. To get links to the textbook and quick-links for the activities in this guide, please visit the resource page for this program:

🖥 https://elementalscience.com/blogs/resources/ps

Two Courses in One

This guide contains the plans for two courses in one book.

- **Hands-on** – The plans in this option are for a standard lab science, one-credit *Physical Science* course. There are textbook assignments, hands-on experiments, and written work with these plans. Expect to take about four to five hours a week to complete these plans.
- **Digital** – The plans in this option are for an online lab science, one-credit *Physical Science* course. There are textbook assignments, online labs, and written work with these plans. Expect to take about four to five hours a week to complete these plans.

Each of the scheduling pages will note at the top which course the plans are for. These schedules for these courses are suggestions; please check with your local oversight contact to make sure that you are meeting your state's graduation requirements. Please feel free to tailor this program to your needs.

An Explanation of the Sections

After the scheduling pages, you will find the notes sheets. These sheets are divided into four sections: textbook, experiments, events in science, and hands-on activities.

Textbook

For this study, we have chosen to use a widely available, standard textbook, *CK-12 Physical Science*. You can download this text as a PDF from the resource page above. You will complete the reading assignment and then answer several of the questions from the text. These answers should be added to the reading section of the science notebook. You will also define several of the key terms from the chapter. The definitions should be added to the glossary section of the science notebook.

Experiment

All the hands-on experiments are in this guide. You will need to have a composition book to record what you do, see, and discover in these experiments. With each of these experiments, you will find the materials, a bit of pre-reading, the procedure, an explanation with the results, and science notebook assignments. If needed, the answers to these assignments will be at the end of the unit materials.

Online Lab

We have also incorporated an optional online lab into the online course. These digital labs are available through Beyond Labz. You can visit the resource page for this program for directions on how to sign up and use these labs or visit their website directly at:

💻 https://www.beyondlabz.com/

Events in Science

This section is designed to familiarize you with current events in science as you research on the internet for the various topics. We have included an article on p. 9 to explain this option in more depth.

Hands-on Activities

We have also included optional hands-on activities for each week.

The Science Notebook

This year, you will create a science notebook. Each notebook should contain the following sections:

- 📖 **Reading** – This section of the notebook will contain any notes you have taken along with the answers to the questions that were assigned each week.
- 📖 **Lab** – This section of the notebook will house the notes from the experiments you have done along with any other materials relating to the labs.
- 📖 **Events** – This section of the notebook will include the current events article summaries you have done.
- 📖 **Glossary** – This section of the notebook will have the definitions for the assigned vocabulary words.

This notebook can be a composition book divided into the required sections or a three-ring binder with dividers for each section.

Final Thoughts

As the authors and publishers of this curriculum, we encourage you to contact us with any questions or problems that you might have concerning *Physical Science* at support@elementalscience.com. We will be more than happy to answer you as soon as we are able. We trust that you and your facilitator will enjoy *Physical Science*.

Adding Current Events into Your Science Studies

Step 1 - Choose the article.

The first step is to choose an appropriate article. Usually, I try to pick one from the field of science that we are studying. You can subscribe to a science magazine, do a Google search, or check out Science News for Students to find possible articles. Once you have collected a list of options, peruse through them, and pick one that you find interesting.

Step 2 - Read the article.

The next step is to read the actual article.

Step 3 - Discuss the article.

The third step is to discuss the article you read. Ideally, your facilitator would ask you the following questions:

- What was the article about?
- What do you think about [a piece of research or an experiment that the article pointed out]?
- How does the article relate to [something that we have studied on the subject]?
- Did you find the article to be interesting?
- Do you agree with the opinion(s) stated in the article?

If your facilitator is not free to discuss the article, answer the above questions on paper.

Step 4 - Write a summary.

The final step to adding current events to your science studies is to write a summary. Once you finish the discussion, write three to five sentences on the article, including your opinion on it. You can use the template on p. 187 as a guide.

A Note to the Facilitator

This guide is written to your students. The idea is that they will work through the materials largely on their own with you being there to help facilitate the class and help them with any questions they may have.

Teacher Edition

There is not a separate teacher guide for *Physical Science*, but you can download a "teacher edition" of the textbook from the resource page for this program:

💻 https://elementalscience.com/blogs/resources/ps

Grading and Credits

The two options in this guide meet the requirements for a full credit of middle school physical science. (**Note**—*You might be able to use this course for high school, but please check with your local oversight group to determine how to award credit.*)

Each week, the students will answer lab and textbook questions, write about events in science, and define vocabulary that can count toward a classwork grade for the course. For the exam grade, the CK-12 textbook for this course has unit tests available for free in the quizzes and tests packet. We suggest you use the following percentages to come up with a final grade for the course:

☞ Class work: 70%
☞ Exam: 30%

We have scheduled the unit tests in the guide. You can use the quizzes at the end of each section as review, or you can count these toward the exam grade as well. A grading rubric for the Scientist Biography Report, which is assigned in Unit 1, can be found on p. 185 in the Appendix.

Physical Science
Unit 1 - Matter

Unit 1 - Overview of Study

Topics Covered

- **Week 1:** What Is Science?
- **Week 2:** Scientific Research, Part 1
- **Week 3:** Scientific Research, Part 2
- **Week 4:** Matter
- **Week 5:** States of Matter
- **Week 6:** Atoms
- **Week 7:** The Periodic Table

Supplies Needed (for the Hands-on Course)

Week	Experiment	Optional Activity
1	*Scientist Study – no supplies needed*	Heavy cream, Milk, Sugar, Vanilla, Crushed ice, 1 Small & 1 Large zip-locking, plastic bag, Dish towel or Oven mitt, Rock salt
2	*Scientist Study – no supplies needed*	*No supplies needed*
3	*Scientist Study – no supplies needed*	Box of toothpicks, Plate, Bowl, Gum-drops
4	4 clear cups, Eye dropper, Table salt, Food coloring, Water	Tall glass, Honey, Corn syrup, Dish soap, Water, Vegetable oil, Rubbing alcohol, Lamp oil
5	Cup, Ice cubes, Pot, Thermometer	Dry ice, Plastic cups, Water, Dish soap
6	*Elemental Profile – no supplies needed*	Pipe cleaners, Pom-pom balls, Glue
7	Periodic Table Match-up Game Templates (*Free download from Elemental Science.*)	*No supplies needed*

Week 1 Notes - What Is Science?

Textbook Assignments
Reading
📖 *CK-12 Physical Science* Sections 1.1, 1.2
Written
After you finish reading, answer questions #1–3, 5, 7 in section 1.1 and questions #1, 3, 5 in section 1.2. File your work in the reading section of your science notebook. Then define the following terms in the glossary section of your science notebook:

- Chemistry
- Induction
- Physical science
- Physics
- Science
- Scientific law
- Scientific theory

Scientist Study
Purpose
The purpose of the scientist study is to learn more about the people who have shaped the field of science. This project will be completed over the next three weeks. For this week, you will complete steps one and two.

Step 1: Choose a Scientist
☞ This week, you need to begin your scientist biography project by choosing which scientist you will research. You can choose one of the scientists mentioned in the textbook or you can choose one that has interested you. Here is a list of famous chemists for more options:
💻 http://chemistry.about.com/od/historyofchemistry/a/famouschemists.htm

Step 2: Research the Scientist
☞ Once you have chosen the scientist you would like to study, you can begin your research. Begin by looking for a biography on your chosen scientist at the library. Then look for articles on the scientist in magazines, newspapers, encyclopedias, or on the internet. You will need to know the following about your scientist to write your report:

- ☑ Biographical information on the scientist (e.g., where they were born, their parents, siblings, how they grew up);
- ☑ The scientist's education (e.g., where they went to school, what kind of student they were, what they studied);
- ☑ Their scientific contributions (e.g., research that they participated in, any significant discoveries they made, the state of the world at the time of their contributions).

As you read over the material you have gathered, be sure to write down any facts you glean in your own words. You can do this on the sheet below or on separate index cards. You will have two weeks to complete your research, so plan accordingly. You

can read more about this method by clicking on the link below:
💻 http://elementalblogging.com/the-index-card-system/

Online Lab
☞ There is no online lab scheduled for this week.

Events in Science
Current Events
🕒 Find a current events article related to the field of chemistry, and complete the article summary sheet found on p. 187 of the Appendix. Once you are done, add the sheet to the events section of your science notebook.

Hands-on Activity
Optional Hands-on
✂ Use a bit of chemistry to make ice cream! You will need heavy cream, milk, sugar, vanilla, crushed ice, one small and one large zip-locking, plastic bag, a dish towel or oven mitt, and rock salt. Mix together the ½ cup cream (120 mL), 1 cup milk (240 mL), ½ cup sugar (225 g) and 1 tsp vanilla (5 mL) in the small zip-locking plastic bag. Then add the crushed ice and rock salt to the large zip-locking plastic bag. (**Note**—*Make sure that the air is removed from both bags and they are sealed tightly*.) Place the smaller bag with the cream mixture inside the larger bag with the ice. Cover the outside of the bag with the dish towel (or put on the oven mitt) and massage or shake the bag until the cream mixture has frozen. It should take about five to ten minutes. Take the smaller bag out of the larger bag, wipe off the salt water, open, eat and enjoy!
- **Troubleshooting Tips:** Use thick plastic bags so that there is less risk of one of the bags being punctured by the ice or salt. If you can't find rock salt, you can use regular table salt instead, but be aware that it may take longer for the cream mixture to freeze.

Week 1	Unit 1 (Hands-on Course)		4-Day
Weekly Topic			
→ This week will be an introduction to science.			

	Day 1	Day 2	Day 3	Day 4
Textbook and Experiment	☐ Read *CK-12 Physical Science* Section 1.1.	☐ Read *CK-12 Physical Science* Section 1.2.	☐ Do the optional hands-on science activity.	
Writing	☐ Add the vocabulary to the glossary section of your science notebook.	☐ Answer the assigned questions in the reading section of your science notebook.	☐ Complete Step 1 of the Scientist Study.	☐ Start Step 2 of the Scientist Study. ☐ Complete the Events in Science assignment.

Supplies Needed

✓ (Optional) Heavy cream, Milk, Sugar, Vanilla, Crushed ice, 1 Small & 1 Large zip-locking, plastic bag, Dish towel or oven mitt, Rock salt

Other Notes

Day 3: Review how to research via the link

Physical Science Unit 1 - Week 1

Week 1 — Unit 1 (Digital Course) — 3-Day

Weekly Topic

→ This week will be an introduction to science.

	Day 1	Day 2	Day 3
Textbook and Experiment	☐ Read *CK-12 Physical Science* Section 1.1.	☐ Read *CK-12 Physical Science* Section 1.2.	☐ *(Optional)* Do the hands-on science activity.
Writing	☐ Add the vocabulary to the glossary section of your science notebook.	☐ Answer the assigned questions in the reading section of your science notebook.	☐ Complete Step 1 and start Step 2 of the Scientist Study. ☐ Complete the Events in Science assignment.

Other Notes

Physical Science Unit 1 - Week 1

Week 2 Notes - Scientific Research, Part 1

Textbook Assignments
Reading
📖 *CK-12 Physical Science* Sections 2.1, 2.2

Written
After you finish reading, answer questions #1–4 in section 2.1 and questions #1–5 in section 2.2. File your work in the reading section of your science notebook. Then define the following terms in the glossary section of your science notebook:

- ☑ Accuracy
- ☑ Control
- ☑ Ethics
- ☑ Hypothesis
- ☑ Kelvin Scale
- ☑ Manipulated Variable
- ☑ Mean
- ☑ Model
- ☑ Observation
- ☑ Precision
- ☑ Range
- ☑ Replication
- ☑ Responding Variable
- ☑ Scientific Notation
- ☑ SI
- ☑ Significant Figures

Scientist Study
Step 2: Research the Scientist
☞ This week, you need to wrap up the research you have done on your scientist. Read over your notes and make sure that you have at least five pieces of information for each of the categories below:
- ☑ Biographical information on the scientist (e.g., where they were born, their parents, siblings, how they grew up);
- ☑ The scientist's education (e.g., where they went to school, what kind of student they were, what they studied);
- ☑ Their scientific contributions (e.g., research that they participated in, any significant discoveries they made, the state of the world at the time of their contributions).

Also, make sure that the recorded list of the resources you have used is complete.

Step 3: Create an Outline
☞ Now that your research is completed, you are ready to begin the process of writing a report on your chosen scientist. This week, you are going to organize the notes you took during step two into a formal outline that you will use next week to write the rough draft of your report. Use the outline template provided on the student sheets as a guide. You should include information on why you chose the particular scientist in your introduction section. For the conclusion section of the outline, you need to include why you believe someone else should learn about your chosen scientist and your impression of the scientist (e.g., Did you like the scientist? Do you feel that they made a significant impact on the field of chemistry?).

Online Lab
☞ There is no online lab scheduled for this week.

Events in Science
Current Events
🕐 Find a current events article related to the field of chemistry, and complete the article summary sheet found on p. 187 of the Appendix. Once you are done, add the sheet to the events section of your science notebook.

Hands-on Activity
Optional Hands-on
✄ There is no optional hands-on activity for this week.

Week 2 — Unit 1 (Hands-on Course) — 4-Day

Weekly Topic

→ This week will be an introduction to scientific research.

	Day 1	Day 2	Day 3	Day 4
Textbook and Experiment	☐ Read *CK-12 Physical Science* Section 2.1.	☐ Read *CK-12 Physical Science* Section 2.2.		
Writing	☐ Add the vocabulary to the glossary section of your science notebook.	☐ Answer the assigned questions in the reading section of your science notebook.	☐ Complete Step 2 of the Scientist Study.	☐ Complete Step 3 of the Scientist Study. ☐ Complete the Events in Science assignment.

Supplies Needed

Other Notes

Week 2	Unit 1 (Digital Course)		3-Day
Weekly Topic			
→ This week will be an introduction to scientific research.			
	Day 1	**Day 2**	**Day 3**
Textbook and Experiment	☐ Read *CK-12 Physical Science* Section 2.1.	☐ Read *CK-12 Physical Science* Section 2.2.	
Writing	☐ Add the vocabulary to the glossary section of your science notebook.	☐ Answer the assigned questions in the reading section of your science notebook.	☐ Complete Steps 2 and 3 of the Scientist Study. ☐ Complete the Events in Science assignment.
Other Notes			

Week 3 Notes - Scientific Research, Part 2

Textbook Assignments
Reading
- *CK-12 Physical Science* Section 2.3

Written
After you finish reading, answer questions #1–4 in section 2.3 and file your work in the reading section of your science notebook. Then define the following terms in the glossary section of your science notebook:
- ☐ Engineer
- ☐ Technology
- ☐ Technological Design

Scientist Study
Step 4: Write a Rough Draft
☞ Last week, you created a formal outline for your scientist biography report; now, it is time to take that outline and turn it into a rough draft. Simply take the points on your outline and combine and add in sentence openers to create a cohesive paragraph. Here's what your rough draft should look like:
- ☑ Paragraph 1 (from outline point I): introduce the scientist;
- ☑ Paragraph 2 (from outline point II): tell about the scientist's education;
- ☑ Paragraph 3–5 (from outline point III): share the scientist's contributions (one paragraph for each contribution); and
- ☑ Paragraph 6 (from outline point IV): share your thoughts on the scientist and why someone should learn about them.

You can choose to hand write or type up your rough draft on a separate sheet of paper. However, keep in mind that you will need a typed version for step five.

Step 5: Revise to Create a Final Draft
☞ Now that you have a typed, double-spaced rough draft, look over it one more time to make any changes you would like. Then have your teacher or one of your peers look over the paper for you to correct any errors and bring clarity to any of the unclear sections. Once this is complete, make the necessary changes to your paper to create your final draft. Print out your final paper, and add it to your science notebook.

Online Lab - Introduction to Scientific Inquiry
Purpose
The purpose of this online lab is to learn about the process of scientific inquiry.

Pre-Reading
☞ Print and read the section of the workbook for the "Introduction to Scientific Inquiry" online lab.

Procedure
- ✓ Do the lab entitled "Introduction to Scientific Inquiry,," and answer the questions as you work through the online lab.

Science Notebook
- ☞ Add the completed workbook pages that you printed to the science notebook.

Events in Science
Current Events
- 🕐 Find a current events article related to the field of chemistry, and complete the article summary sheet found on p. 187 of the Appendix. Once you are done, add the sheet to the events section of your science notebook.

Hands-on Activity
Optional Hands-on
- ✂ Engineer a gumdrop tower! You will need a box of toothpicks, a plate, a bowl, and gumdrops. Set out the gumdrops in a bowl, the box of toothpicks on the table, and a plate to build on. Then design, test, and build whatever gumdrop tree you can dream up.

Week 3 — Unit 1 (Hands-on Course) — 4-Day

Weekly Topic

→ This week will continue the introduction to scientific research.

	Day 1	Day 2	Day 3	Day 4
Textbook and Experiment	☐ Read *CK-12 Physical Science* Section 2.3.		☐ Do the optional hands-on science activity.	
Writing	☐ Add the vocabulary to the glossary section of your science notebook.	☐ Answer the assigned questions in the reading section of your science notebook.	☐ Complete Step 4 of the Scientist Study.	☐ Complete Step 5 of the Scientist Study. ☐ Complete the Events in Science assignment.

Supplies Needed

✓ *(Optional)* Box of toothpicks, Plate, Bowl, Gumdrops

Other Notes

Week 3 — Unit 1 (Digital Course) — 3-Day

Weekly Topic

→ This week will continue the introduction to scientific research.

	Day 1	Day 2	Day 3
Textbook and Experiment	☐ Read *CK-12 Physical Science* Section 2.3.	☐ Do the online lab "Introduction to Scientific Inquiry."	☐ *(Optional)* Do the hands-on science activity.
Writing	☐ Add the vocabulary to the glossary section of your science notebook.	☐ Answer the assigned questions in the reading section of your science notebook.	☐ Complete Steps 4 and 5 of the Scientist Study. ☐ Complete the Events in Science assignment.

Other Notes

Week 4 Notes - Matter

Textbook Assignments
Reading
📖 *CK-12 Physical Science* Sections 3.1, 3.2, 3.3

Written
After you finish reading, answer questions #1–5 in section 3.1, questions #2–6 in section 3.2, and questions #2–6 in section 3.3. File your work in the reading section of your science notebook. Then define the following terms in the glossary section of your science notebook:

- Chemical Change
- Chemical Property
- Colloid
- Crystal
- Density
- Law of Conservation of Mass
- Mixture
- Physical Change
- Reactivity
- Suspension
- Volume

Experiment - How does the addition of salt affect the density of water?

Materials
- 4 clear cups
- Eye dropper
- Table salt
- Food coloring
- Water

Pre-Reading
Density is a property of matter that relates mass and volume. Mass tells you how much matter is in something, whereas volume tells you how much space something occupies. In other words, density measures how much matter of a particular substance will occupy a given space. When comparing two different solutions, a denser solution will sink to the bottom while a less dense solution will float to the top. In this week's experiment, you are going to use the principle of density to see how salt affects the density of water.

Procedure

1. Read the introduction to this experiment and make a hypothesis. Then prepare your solutions according to the directions below.
   - Begin by putting 1 cup (240 mL) of room temperature water into each of the 4 bowls.
   - For the yellow solution, add 10 drops of yellow food coloring to one of the bowls.
   - For the red solution, add 10 drops of red food coloring and 2 tbsp (34.2 g) of salt to another bowl.
   - For the blue solution, add 10 drops of blue food coloring and ¼ cup (68.4 g) of salt to another bowl.
   - For the green solution, add 10 drops of green food coloring and 6 tbsp (102.6 g) of salt to the last bowl.

↳ Stir each one to make sure that the salt is thoroughly dissolved.
2. Label each of your cups #1–4. Fill cup #1 with 1 cup (240 mL) of the yellow solution, cup #2 with 1 cup (240 mL) of the red solution, cup #3 with 1 cup (240 mL) of the blue solution, and cup #4 with 1 cup (240 mL) of the green solution.
3. Using an eye dropper, draw up about 1–2 tsp (5–10 mL) of the yellow solution and *very slowly* add it to the cup #2. (**Note**—*When you add the yellow solution, you need to make sure the tip of your eyedropper is just below the surface of the solution in the cup.*) Wait 1 minute and record your observations.
4. Next, using an eye dropper, draw up about 1–2 tsp (5–10 mL) of the yellow solution and *very slowly* add it to the cup #3. Wait 1 minute, and record your observations.
5. Then, using an eye dropper, draw up about 1–2 tsp (5–10 mL) of the yellow solution and *very slowly* add it to the cup #4. Wait 1 minute, and record your observations.
6. Pour out the solutions from each of the used cups and refill them with 1 cup of each of the colors as before. Repeat steps 3–6 with the red, blue, and green solutions.

Explanation
The addition of salt to the water increases the overall mass of the solution but minimally affects the volume. This means that the density will increase. The more salt you add to the water, the denser the solution will become. This is the reason why you saw that the yellow liquid always floated to the top of each cup, whereas the green liquid always sank to the bottom.

Science Notebook
☞ Write down a short description of what you did for the experiment and add a chart with results. After that, calculate the density of the four solutions you made. (See Appendix pp. 188–189 for more information about calculating density.) Finally, add what you have learned from the experiment.

Online Lab - Rutherford's Experiment and the Nucleus
Purpose
The purpose of this online lab is to learn how Rutherford found evidence for the nucleus of the atom.
Pre-Reading
✐ Print and read the section of the workbook for the "Rutherford's Experiment and the Nucleus" online lab.
Procedure
✓ Do the lab entitled "Rutherford's Experiment and the Nucleus,," and answer the questions as you work through the online lab.
Science Notebook
☞ Add the completed workbook pages that you printed to the science notebook.

Events in Science
Current Events
🕐 Find a current events article related to the field of chemistry, and complete the article summary sheet found on p. 187 of the Appendix. Once you are done, add the sheet to the events section of your science notebook.

Hands-on Activity
Optional Hands-on
✂ Make a density column. You will need ½ cup (240 mL) of honey, corn syrup, dish soap, water, vegetable oil, rubbing alcohol, and lamp oil. Add a different shade of food coloring to each of the liquids except the honey and vegetable oil. (**Note**—*If your dish soap is already colored, you do not need to add a color to that.*) Next, pour the honey into a cylindrical container. Then slowly add the corn syrup on top. You want to pour the liquid slowly down the side of the container to minimize the bubbles and give you the best results. Follow with the dish soap, water, vegetable oil, alcohol, and finally the lamp oil. You should have seven separate liquid layers on top of each other. Each layer has the same volume but different densities.

Week 4 — Unit 1 (Hands-on Course) — 4-Day

Weekly Topic

→ This week will begin a look at matter.

	Day 1	Day 2	Day 3	Day 4
Textbook and Experiment	☐ Read *CK-12 Physical Science* Section 3.1 and 3.2.	☐ Read *CK-12 Physical Science* Section 3.3.	☐ Do the experiment "How does the addition of salt affect the density of water?"	☐ Do the optional hands-on science activity.
Writing	☐ Add the vocabulary to the glossary section of your science notebook.	☐ Answer the assigned questions in the reading section of your science notebook.	☐ Record what you have done in the lab section of your science notebook.	☐ Complete the Events in Science assignment.

Supplies Needed

✓ **Experiment** – 4 clear cups, Eye dropper, Table salt, Food coloring, Water
✓ *(Optional)* Tall glass, Honey, Corn syrup, Dish soap, Water, Vegetable oil, Rubbing alcohol, Lamp oil

Other Notes

Week 4 — Unit 1 (Digital Course) — 3-Day

Weekly Topic

→ This week will begin a look at matter.

	Day 1	Day 2	Day 3
Textbook and Experiment	☐ Read *CK-12 Physical Science* Section 3.1 and 3.2.	☐ Read *CK-12 Physical Science* Section 3.3.	☐ Do the online lab "Rutherford's Experiment and the Nucleus." ☐ *(Optional)* Do the hands-on science activity.
Writing	☐ Add the vocabulary to the glossary section of your science notebook.	☐ Answer the assigned questions in the reading section of your science notebook.	☐ Record what you have done in the lab section of your science notebook. ☐ Complete the Events in Science assignment.

Other Notes

Week 5 Notes - States of Matter

Textbook Assignments
Reading
- 📖 *CK-12 Physical Science* Section 4.1, 4.2, 4.3

Written
After you finish reading, answer questions #1, 3, 5, 7 in section 4.1, questions #1, 3, 5 in section 4.2, and questions #1, 3, 5, 7 in section 4.3. File your work in the reading section of your science notebook. Then define the following terms in the glossary section of your science notebook:

- ☐ Amontons' Law
- ☐ Boyle's Law
- ☐ Charles's Law
- ☐ Condensation
- ☐ Energy
- ☐ Evaporation
- ☐ Kinetic Energy
- ☐ Kinetic Theory of Matter
- ☐ Plasma
- ☐ States of Matter
- ☐ Sublimation
- ☐ Vaporization

Experiment - When does water change state?
Materials
- ✓ Cup
- ✓ Ice cubes
- ✓ Pot
- ✓ Thermometer

Pre-Reading
There are three states of matter that are easily observed every day: solid, liquid, and gas. Solids have tightly packed molecules with a fixed shape and volume; liquids have widely spaced molecules with a fixed volume; and gases have independently moving molecules with no fixed shape or volume. There is also a fourth state of matter called plasma, which is rarely seen on Earth. Plasma is created when a gas ionizes at very high temperatures. In this experiment, you are going to look at the points at which water changes state.

Procedure
1. Fill a small pot halfway with ice cubes. Place the pot on a burner and turn the burner onto medium heat.
    - **Caution:** Do not touch the burner! Be sure to use protective gear when working near the burner.
2. Observe the thermometer as the ice begins to melt and record the temperature once all of the ice melts.
3. Continue to heat the water, observing the temperature on the thermometer as it heats up.
4. Once you begin to see the water boiling and observe the presence of steam, record your last temperature measurement.
5. Turn the burner off and remove the pot from the burner before you draw conclusions.

Troubleshooting Tips
If the temperature goes way up, but your water is not boiling, make sure that your thermometer is not touching the bottom of the pan. If it is, you will get a false reading.

Also, your elevation can effect your results. The melting point and boiling point of water are given at one atmosphere of pressure, which is at sea level. If you live above or below sea level, this will affect the temperature at which your ice melts and your water boils. The effect will depend upon how much you are above or below sea level. Generally, the higher your elevation, the higher your temperatures will be.

Explanation
The melting point of water is 32° F (0° C), which is the point at which it changes from a solid to a liquid. This is the reason that you began to see the ice melting to liquid water around that temperature. The boiling point of water is 212° F (100° C), which is the point at which all the water changes from a liquid to a gas. You will have stopped the experiment before this point so that you did not burn the pan. However, if we let the experiment carry on, you would reach the boiling point of water when all the liquid water had turned to steam.

Science Notebook
☞ Write down a short description of what you did for the experiment and add a chart with results. After that, add what you have learned from the experiment.

Online Lab - Temperature During Melting
Purpose
The purpose of this online lab is to determine how temperature changes as ice is heated and becomes liquid water.

Pre-Reading
✎ Print and read the section of the workbook for the "Temperature During Melting" online lab.

Procedure
✓ Do the lab entitled "Temperature During Melting," and answer the questions as you work through the online lab.

Science Notebook
☞ Add the completed workbook pages that you printed to the science notebook.

Events in Science
Current Events
🕐 Find a current events article related to the field of chemistry, and complete the article summary sheet found on p. 187 of the Appendix. Once you are done, add the sheet to the events section of your science notebook.

Hands-on Activity
Optional Hands-on
✂ Dry ice is the solid form of carbon dioxide. It's a fun chemical to play around with because it readily sublimes (changes from solid to gas) at atmospheric pressure. Check out the following article for activity suggestions and a coordinating worksheet. You will need dry ice, water, plastic cups, and dish soap for these activities.
💻 http://elementalblogging.com/dry-ice-exploration/

Week 5 — Unit 1 (Hands-on Course) — 4-Day

Weekly Topic

→ This week will look at states of matter

	Day 1	Day 2	Day 3	Day 4
Textbook and Experiment	☐ Read *CK-12 Physical Science* Section 4.1 and 4.2.	☐ Read *CK-12 Physical Science* Section 4.3.	☐ Do the experiment "When does water change state?"	☐ Do the optional hands-on science activity.
Writing	☐ Add the vocabulary to the glossary section of your science notebook.	☐ Answer the assigned questions in the reading section of your science notebook.	☐ Record what you have done in the lab section of your science notebook.	☐ Complete the Events in Science assignment.

Supplies Needed

- ✓ **Experiment** – Cup, Ice cubes, Pot, Thermometer
- ✓ *(Optional)* Dry ice, Plastic cups, Water, Dish soap

Other Notes

Week 5 — Unit 1 (Digital Course) — 3-Day

Weekly Topic

→ This week will begin a look at matter.

	Day 1	Day 2	Day 3
Textbook and Experiment	☐ Read *CK-12 Physical Science* Section 4.1 and 4.2.	☐ Read *CK-12 Physical Science* Section 4.3.	☐ Do the online lab "Temperature During Melting." ☐ *(Optional)* Do the hands-on science activity.
Writing	☐ Add the vocabulary to the glossary section of your science notebook.	☐ Answer the assigned questions in the reading section of your science notebook.	☐ Record what you have done in the lab section of your science notebook. ☐ Complete the Events in Science assignment.

Other Notes

Week 6 Notes - Atoms

Textbook Assignments
Reading
📖 *CK-12 Physical Science* Sections 5.1, 5.2, 5.3

Written
After you finish reading, answer questions #1, 3, 5, 7 in section 5.1, questions #1–6 in section 5.2, and questions #1–5 in section 5.3. File your work in the reading section of your science notebook. Then define the following terms in the glossary section of your science notebook:

- ☐ Atomic Mass Unit
- ☐ Electron
- ☐ Electron Cloud
- ☐ Energy Level
- ☐ Isotope
- ☐ Orbital
- ☐ Proton

Element Profile Page
This week you will spend time researching an element of your choice and creating a profile page for that element.

Steps to Follow:
1. Begin by choosing one of the elements of the periodic table for an in-depth profile.
2. Do some research to find out more about your chosen element. Use the internet and the resources you have in your home or at your library to find out more about the element.
3. Answer the following questions from your research:
   - What are the name, symbol, atomic number, and atomic mass of the element?
   - Who discovered the element, and when did that happen?
   - Is it a gas, liquid, or solid at room temperature?
   - Where is the element typically found?
   - What are the major uses of the element?
4. Be sure to write down any additional interesting facts you have learned on individual index cards.
5. Finally, complete the profile page for your element. Fill out the top left-hand rectangle with the information on the element from the periodic table. Then answer the questions, and write a brief summary on the element and its uses. You may also want to include the story of how the element was discovered in your summary. (Get a template for this profile page on p. 190 of the Appendix.)

Online Lab - Temperature During Vaporization
Purpose
The purpose of this online lab is to investigate how the temperature of liquid water changes as it is heated until it becomes water vapor.

Pre-Reading
↪ Print and read the section of the workbook for the "Temperature During Vaporization" online lab.

Procedure
✓ Do the lab entitled "Temperature During Vaporization," and answer the questions as you work through the online lab.

Science Notebook
☞ Add the completed workbook pages that you printed to the science notebook.

Events in Science
Current Events
🕒 Find a current events article related to the field of chemistry, and complete the article summary sheet found on p. 187 of the Appendix. Once you are done, add the sheet to the events section of your science notebook.

Hands-on Activity
Optional Hands-on
✂ Make a model of an atom using pipe cleaners, pom-pom balls, and glue. They can use red balls for protons (p), brown balls for neutrons (n), and yellow balls for electrons (e). Have them glue the protons and neutrons together for a nucleus. Then glue the electrons onto yellow pipe cleaners and make a circle around the nucleus. Have the students try to make a model of helium (2p, 2n, 2e) or carbon (6p, 6n, 6e).

Week 6 — Unit 1 (Hands-on Course) — 4-Day

Weekly Topic

→ This week will look at the atom.

	Day 1	Day 2	Day 3	Day 4
Textbook and Experiment	☐ Read *CK-12 Physical Science* Section 2.3.		☐ Do the optional hands-on science activity.	
Writing	☐ Add the vocabulary to the glossary section of your science notebook.	☐ Answer the assigned questions in the reading section of your science notebook.	☐ Begin working on the Element Profile Page.	☐ Complete the Element Profile Page. ☐ Complete the Events in Science assignment.

Supplies Needed

✓ *(Optional)* Pipe cleaners, Pom-pom balls, Glue

Other Notes

Week 6 — Unit 1 (Digital Course) — 3-Day

Weekly Topic

→ This week will look at the atom.

	Day 1	Day 2	Day 3
Textbook and Experiment	❐ Read *CK-12 Physical Science* Section 2.3.	❐ Do the online lab "Temperature During Vaporization."	❐ *(Optional)* Do the hands-on science activity.
Writing	❐ Add the vocabulary to the glossary section of your science notebook.	❐ Answer the assigned questions in the reading section of your science notebook.	❐ Complete the Element Profile Page. ❐ Complete the Events in Science assignment.

Other Notes

Week 7 Notes - The Periodic Table

Textbook Assignments
Reading
📖 *CK-12 Physical Science* Sections 6.1, 6.2, 6.3

Written
After you finish reading, answer questions #1–5 in section 6.1, questions #1, 3, 5, 7 in section 6.2, and questions #1, 3, 5, 7 in section 6.3. File your work in the reading section of your science notebook. Then define the following terms in the glossary section of your science notebook:

- ☐ Alkali Metal
- ☐ Alkaline Earth Metal
- ☐ Group
- ☐ Halogen
- ☐ Metal
- ☐ Metalloid
- ☐ Nonmetal
- ☐ Period
- ☐ Periodic Table
- ☐ Transition Metal
- ☐ Valence Electron

Unit Test
Take Unit Test #1 – Science and Technology and Unit Test #2 – Matter starting on *CK-12 Physical Science*, p. 336.

Game - The Periodic Table Match-up Game
Purpose
The purpose of playing this game is to help you work on memorizing the elements in the periodic table.

Pre-Reading
✍ Watch the following video to begin memorizing the elements.
   💻 https://www.youtube.com/watch?v=rz4Dd1I_fX0

Procedure
✓ Download the following game and follow the directions to play Periodic Table Match-up.
   💻 https://elementalscience.com/products/chemistry-game-periodic-table-match-up-free-ebook

Online Lab - Elements and the Periodic Table
Purpose
The purpose of this online lab is to find out what information about the elements can be obtained from the periodic table.

Pre-Reading
✍ Print and read the section of the workbook for the "Elements and the Periodic Table"

online lab.

Procedure
✓ Do the lab entitled "Elements and the Periodic Table," and answer the questions as you work through the online lab.

Science Notebook
☞ Add the completed workbook pages that you printed to the science notebook.

Events in Science
Current Events
🕒 Find a current events article related to the field of chemistry, and complete the article summary sheet found on p. 187 of the Appendix. Once you are done, add the sheet to the events section of your science notebook.

Hands-on Activity
Optional Hands-on
✂ Create a periodic table to display on the wall. You can purchase a visual representation of the periodic table from one of the following sites to use as a guide:
💻 http://elements.wlonk.com/ (You can download a 8 ½ x 11 pdf version for free.)
💻 http://periodictable.com/ (This is my personal favorite, but it is pricey.)

I have included a blank periodic table in the Appendix on p. 191 that you may increase in size to use on the wall.

Week 7 — Unit 1 (Hands-on Course) — 4-Day

Weekly Topic

→ This week will look at the periodic table.

	Day 1	Day 2	Day 3	Day 4
Textbook and Experiment	☐ Read *CK-12 Physical Science* Section 2.3.	☐ Play the Periodic Table Match-up game.	☐ Do the optional hands-on science activity.	☐ Take Unit Test #1 – Science and Technology and Unit Test #2 – Matter starting on *CK-12 Physical Science*, p. 336.
Writing	☐ Add the vocabulary to the glossary section of your science notebook.	☐ Answer the assigned questions in the reading section of your science notebook.		☐ Complete the Events in Science assignment.

Supplies Needed

✓ **Experiment** – Periodic Table Match-up game templates (*free download from Elemental Science*)

Other Notes

Week 7 — Unit 1 (Digital Course) — 3-Day

Weekly Topic

→ This week will look at the periodic table.

	Day 1	Day 2	Day 3
Textbook and Experiment	☐ Read *CK-12 Physical Science* Section 2.3.	☐ Do the online lab "Elements of the Periodic Table."	☐ *(Optional)* Do the hands-on science activity. ☐ Take Unit Test #1 – Science and Technology and Unit Test #2 – Matter starting on *CK-12 Physical Science*, p. 336.
Writing	☐ Add the vocabulary to the glossary section of your science notebook.	☐ Answer the assigned questions in the reading section of your science notebook.	☐ Complete the Events in Science assignment.

Other Notes

Unit 1 - Answers

Week 4
Experiment
- Your results chart should look like this:

	Added to the Yellow Solution	Added to the Red Solution	Added to the Blue Solution	Added to the Green Solution
Yellow Solution		Floated	Floated	Floated
Red Solution	Sank		Floated	Floated
Blue Solution	Sank	Sank		Floated
Green Solution	Sank	Sank	Sank	

Science Notebook Assignment (Experiment)
- Yellow solution = 1.00 g/mL
- Red solution = 1.14 g/mL
- Blue solution = 1.29 g/mL
- Green solution = 1.43 g/mL

Week 5
Experiment
- You should see that the water began to melt at 32° F (0° C) and that it began to boil around 212° F (100° C).

Physical Science

Unit 2 – Chemical Properties

Physical Science Unit 2

Unit 2 - Overview of Study

Topics Covered

- **Week 1:** Chemical Bonding, Part 1
- **Week 2:** Chemical Bonding, Part 2
- **Week 3:** Chemical Reactions, Part 1
- **Week 4:** Chemical Reactions, Part 2
- **Week 5:** Chemistry of Carbon, Part 1
- **Week 6:** Chemistry of Carbon, Part 2
- **Week 7:** Chemistry of Solutions
- **Week 8:** Nuclear Chemistry

Supplies Needed (for the Hands-on Course)

Week	Experiment	Optional Activity
1	Cake frosting, Red and yellow bite-sized candies (such as regular-sized M&Ms)	*No supplies needed*
2	Cake frosting, Red and yellow bite-sized candies (such as regular-sized M&Ms)	*No supplies needed*
3	Baking soda, Chalk, Iron nail (non-coated), Copper penny, Aluminum foil, White vinegar, 5 cups	*No supplies needed*
4	Yeast, Hydrogen peroxide, Epsom salts, Water, 2 Cups, 2 Thermometers	Hydrogen peroxide, Dish soap, Yeast, Water, Bottle, Cup
5	Sugar, Salt, Candle, 2 Metal spoons, Hot mitt	Vegetable oil, Margarine
6	2 Slices of bread, Water, Saliva, 2 Plastic bags	An iodine swab, Several different types of food for testing (such as a hard-boiled egg, bread, potato, pasta, yogurt, cookies, or cheese)
7	5 Clear cups (or beakers), 5 Plastic spoons, Sugar, Salt, Baking soda, Flour, Petroleum jelly, Water, Vegetable oil, Tablespoon	A head of red cabbage, Water, Pot, A variety of liquids or powders from your kitchen to test, such as lemon juice, baking soda, soda, or detergent, Several cups
8	*No supplies needed*	*No supplies needed*

Week 1 Notes - Chemical Bonding, Part 1

Textbook Assignments

Reading
📖 *CK-12 Physical Science* Sections 7.1, 7.2

Written
After you finish reading, answer questions #1–4 in section 7.1 and questions #1–6 in section 7.2. File your work in the reading section of your science notebook. Then define the following terms in the glossary section of your science notebook:

- Chemical Bond
- Chemical Formula
- Ionic Bond
- Ionic Compound

Experiment - Ionic Bonding

Materials
- ✓ Cake frosting
- ✓ Red and yellow bite-sized candies (such as regular-sized M&Ms)

Pre-Reading
There are three main ways that atoms can chemically bond to form molecules: ionic bonding, covalent bonding, and metallic bonding. This week, you are going to look closer at ionic bonding. In this type of bonding, one atom donates an electron to another to form the bond. In this experiment, you are going to use candy to demonstrate this type of chemical bonding. After that, you are going to practice naming ionic compounds in your science notebook.

Procedure
1. Count out nine red candies and one yellow candy.
2. Arrange the nine red candies in a circle, so that there are two in the center with seven surrounding them. (Be sure to leave a space for one more in the outer circle.) These are the electrons in the outer shell of a chlorine atom.
3. Take the one yellow candy, which represents the electron in the outer shell of a sodium atom, and move it into the space in the outer circle of the chlorine electrons. Now you have an ionic bond between your sodium and chlorine atom.

Explanation
This experiment is meant to be a demonstration of how ionic bonding works so that you can picture it in your mind. To see a picture of what your creation should look like, please

check out the unit answer sheet.

Science Notebook
- ☞ Draw a picture of your ionic chocolate bond, being sure to label the atoms.
- ☞ Complete the ionic compound nomenclature practice by reading about nomenclature from the worksheet in the Appendix on p. 192. Then name the ionic compounds on p. 193.

Online Lab - Creating Chemical Compounds

Purpose
The purpose of this online lab is to determine the ration of elements in a chemical compound.

Pre-Reading
- ✍ Print and read the section of the workbook for the "Creating Chemical Compounds" online lab.

Procedure
- ✓ Do the lab entitled "Creating Chemical Compounds," and answer the questions as you work through the online lab.

Science Notebook
- ☞ Add the completed workbook pages that you printed to the science notebook.

Events in Science

Current Events
- ⏲ Find a current events article related to the field of chemistry, and complete the article summary sheet found on p. 187 of the Appendix. Once you are done, add the sheet to the events section of your science notebook.

Hands-on Activity

Optional Hands-on
- ✂ Think of some changes that you see every day, such as water being boiled or making a cup of tea. List these changes and then decide if they are chemical or physical changes. (**Note**—*Remember that a chemical change looks different and it is usually a permanent irreversible change. A physical change can look different, but chemically it is still the same. It is not permanent and can be reversed.*) Once you have at least five of each, create a poster depicting the difference between chemical and physical changes.

Week 1	Unit 2 (Hands-on Course)			4-Day
Weekly Topic				
→ This week will be an introduction to chemical bonding.				
	Day 1	**Day 2**	**Day 3**	**Day 4**
Textbook and Experiment	❏ Read *CK-12 Physical Science* Section 7.1.	❏ Read *CK-12 Physical Science* Section 7.2.	❏ Do the experiment "Ionic Bonding."	
Writing	❏ Add the vocabulary to the glossary section of your science notebook.	❏ Answer the assigned questions in the reading section of your science notebook.	❏ Record what you have done in the lab section of your science notebook.	❏ Complete the Events in Science assignment.
Supplies Needed				
✓ **Experiment** – Cake frosting, Red and yellow bite-sized candies (such as regular-sized M&Ms)				
Other Notes				

Physical Science Unit 2 - Week 1

Week 1	Unit 2 (Digital Course)		3-Day
Weekly Topic			
→ This week will be an introduction to chemical bonding.			
	Day 1	**Day 2**	**Day 3**
Textbook and Experiment	☐ Read *CK-12 Physical Science* Section 7.1.	☐ Read *CK-12 Physical Science* Section 7.2.	☐ Do the online lab "Creating Chemical Compounds."
Writing	☐ Add the vocabulary to the glossary section of your science notebook.	☐ Answer the assigned questions in the reading section of your science notebook.	☐ Record what you have done in the lab section of your science notebook. ☐ Complete the Events in Science assignment.
Other Notes			

Physical Science Unit 2 - Week 1

Week 2 Notes - Chemical Bonding, Part 2

Textbook Assignments

Reading
📖 *CK-12 Physical Science* Sections 7.3, 7.4

Written
After you finish reading, answer questions #1, 3, 5 in section 7.3 and questions #1–4 in section 7.4. File your work in the reading section of your science notebook. Then define the following terms in the glossary section of your science notebook:

- ☐ Alloy
- ☐ Covalent Bond
- ☐ Hydrogen Bond
- ☐ Metallic Bond
- ☐ Nonpolar
- ☐ Polar

Experiment - Covalent Bonding

Materials
- ✓ Cake frosting
- ✓ Red and yellow bite-sized candies (such as regular-sized M&Ms)

Pre-Reading
As we learned last week, there are three main ways that atoms can chemically bond to form molecules: ionic bonding, covalent bonding, and metallic bonding. This week, you are going to look closer at covalent bonding. In this type of bonding, one atom shares an electron with another atom to form the bond. In this experiment, you are going to use candy to demonstrate this type of chemical bonding. After that, you are going to practice naming covalent compounds in your science notebook.

Procedure
1. Count out eight red candies and two yellow candies.
2. Use the eight red candies to make a model showing the electrons in the outer shell of an oxygen atom by making a circle with six candies on the outside and two on the inside. (Be sure to leave space for two more candies in the outer layer.)
3. Now, take two of the yellow candies, which represent the electrons in the outer shells of the two different hydrogen atoms, and bite or cut them in half.
4. Do the same with the two of the red candies from the outer circle you created.
5. Then use the frosting to glue together one half of each color to form four whole candies.
6. Place the dual-colored candies in the outer circle of your model to represent the covalent bond between oxygen and hydrogen in water.

Explanation
This experiment is meant to demonstrate of how covalent bonding works so that you can

picture it in your mind. To see a picture of what your creation should look like, please check out the unit answer sheet.

Science Notebook
- Draw a picture of your covalent chocolate bond, being sure to label the atoms.
- Complete the covalent compound nomenclature practice by reading about nomenclature from the worksheet in the Appendix on p. 192. Then name the covalent compounds on p. 193.

Online Lab
- There is no online lab scheduled for this week.

Events in Science
Current Events
- Find a current events article related to the field of chemistry, and complete the article summary sheet found on p. 187 of the Appendix. Once you are done, add the sheet to the events section of your science notebook.

Hands-on Activity
Optional Hands-on
- There is no optional hands-on activity for this week.

Week 2	Unit 2 (Hands-on Course)			4-Day
Weekly Topic				
→ This week will wrap up a look at chemical bonding.				
	Day 1	Day 2	Day 3	Day 4
Textbook and Experiment	☐ Read *CK-12 Physical Science* Section 7.3.	☐ Read *CK-12 Physical Science* Section 7.4.	☐ Do the experiment "Covalent Bonding"	☐ Do the optional hands-on science activity.
Writing	☐ Add the vocabulary to the glossary section of your science notebook.	☐ Answer the assigned questions in the reading section of your science notebook.	☐ Record what you have done in the lab section of your science notebook.	☐ Complete the Events in Science assignment.

Supplies Needed

✓ **Experiment** – Cake frosting, Red and yellow bite-sized candies (such as regular-sized M&Ms)
✓ *(Optional)* 5 Clear cups (or beakers), 5 Plastic spoons, Sugar, Salt, Baking soda, Flour, Petroleum jelly, Water, Vegetable oil, and a Tablespoon

Other Notes

Week 2	Unit 2 (Digital Course)		3-Day
Weekly Topic			
→ This week will wrap up a look at chemical bonding.			
	Day 1	Day 2	Day 3
Textbook and Experiment	☐ Read *CK-12 Physical Science* Section 7.3.	☐ Read *CK-12 Physical Science* Section 7.4.	☐ *(Optional)* Do the hands-on science activity.
Writing	☐ Add the vocabulary to the glossary section of your science notebook.	☐ Answer the assigned questions in the reading section of your science notebook.	☐ Complete the Events in Science assignment.
Other Notes			

Week 3 Notes - Chemical Reactions, Part 1

Textbook Assignments

Reading
📖 *CK-12 Physical Science* Sections 8.1, 8.2

Written
After you finish reading, answer questions #1–5 in section 8.1 and questions #1–3 in section 8.2. File your work in the reading section of your science notebook. Then define the following terms in the glossary section of your science notebook:

- ☐ Chemical Equation
- ☐ Equilibrium
- ☐ Product
- ☐ Reactant

Experiment - How reactive is it?

Materials
- ✓ Baking soda
- ✓ Chalk
- ✓ Iron nail (non-coated)
- ✓ Copper penny
- ✓ Aluminum foil
- ✓ White vinegar
- ✓ 5 cups

Pre-Reading
The reactivity series tells how reactive a given metal can be in its elemental state. Those at the top of the reactivity series, such as potassium, change quickly when exposed to air, while those at the bottom, such as gold, do not change at all. We can also use the periodic table to help us determine an element's propensity to react. For metals, reactivity decreases as you move from left to right on the periodic table and increases as you move down the table. In this experiment, you are going to test the reactivity of sodium (baking soda), calcium (chalk), iron (nail), aluminum (foil), and copper (penny) in vinegar.

Procedure
1. Pour ½ cup (120 mL) of vinegar into each of the 5 cups.
2. Add 1 tsp (3.5 g) of baking soda to the first cup. Observe and record how long it takes for a reaction to occur.
3. Repeat this process with 1 piece of chalk, 1 iron nail, a piece of aluminum foil, and 1 copper penny in the other 4 cups.
4. Observe and record how long it takes for a reaction to occur.

Explanation
You should see that the baking soda reacted immediately and vigorously. The chalk

should have begun to bubble lightly fairly soon, but it took hours to dissolve fully. You should also see that the iron nail begins to rust after several hours. The aluminum foil should eventually turn grayish-black. Finally, the penny should have shown little to no change after several hours in the vinegar.

Sodium is very high on the reactivity series, which explains why the baking soda (sodium bicarbonate) reacted quickly with the vinegar. Calcium is still relatively high on the series, but not nearly as much as sodium. This is why you saw a reaction beginning right away, but that reaction took much longer to complete than the one with sodium. Aluminum is below calcium, but above iron, which is why this reaction falls in between the other two. Iron is even further down on the reactivity series, which explains why the nail took longer to rust. Finally, copper is the lowest metal in the series of the elements we tested, which explains why it took the longest to react.

Science Notebook
- Write down a short description of what you did for the experiment and add a chart with results. After that, add what you have learned from the experiment.
- Complete the balancing chemical equations practice by reading about balancing equations from the worksheet in the Appendix on pp. 194–195.

Online Lab
- There is no online lab scheduled for this week.

Events in Science

Current Events
- Find a current events article related to the field of chemistry, and complete the article summary sheet found on p. 187 of the Appendix. Once you are done, add the sheet to the events section of your science notebook.

Hands-on Activity

Optional Hands-on
- Watch an equilibrium reaction that is affected by temperature:
  - Equilibrium in Nitrogen Dioxide Gas – https://www.youtube.com/watch?v=zVZXq64HSV4

Week 3 — Unit 2 (Hands-on Course) — 4-Day

Weekly Topic

→ This week will be an introduction to chemical reactions.

	Day 1	Day 2	Day 3	Day 4
Textbook and Experiment	☐ Read *CK-12 Physical Science* Section 8.1.	☐ Read *CK-12 Physical Science* Section 8.2.	☐ Do the experiment "How reactive is it?"	☐ Do the optional hands-on science activity.
Writing	☐ Add the vocabulary to the glossary section of your science notebook.	☐ Answer the assigned questions in the reading section of your science notebook.	☐ Record what you have done in the lab section of your science notebook.	☐ Complete the Events in Science assignment.

Supplies Needed

✓ **Experiment** – Baking soda, Chalk, Iron nail (non-coated), Copper penny, Aluminum foil, White vinegar, 5 Cups

Other Notes

Week 3	Unit 2 (Digital Course)		3-Day
Weekly Topic			
→ This week will be an introduction to chemical reactions.			
	Day 1	**Day 2**	**Day 3**
Textbook and Experiment	☐ Read *CK-12 Physical Science* Section 8.1.	☐ Read *CK-12 Physical Science* Section 8.2.	☐ *(Optional)* Do the hands-on science activity.
Writing	☐ Add the vocabulary to the glossary section of your science notebook.	☐ Answer the assigned questions in the reading section of your science notebook.	☐ Complete the Events in Science assignment.
Other Notes			

Week 4 Notes - Chemical Reactions, Part 2

Textbook Assignments

Reading
📖 *CK-12 Physical Science* Section 8.3, 8.4

Written
After you finish reading, answer questions #2, 4, 6 in section 8.3 and questions #2, 4, 6, 8 in section 8.4. File your work in the reading section of your science notebook. Then define the following terms in the glossary section of your science notebook:

- ☐ Activation Energy
- ☐ Combustion Reaction
- ☐ Decomposition Reaction
- ☐ Endothermic Reaction
- ☐ Exothermic Reaction
- ☐ Law of Conservation of Energy
- ☐ Replacement Reaction
- ☐ Synthesis Reaction

Experiment - Exothermic or Endothermic

Materials
- ✓ Yeast
- ✓ Hydrogen peroxide
- ✓ Epsom salts
- ✓ Water
- ✓ 2 Cups
- ✓ 2 Thermometers

Pre-Reading
In a chemical reaction, bonds are broken and reformed. This causes the reactants, or the substances you begin with, to be turned into new substances known as the products. Every chemical reaction involves energy that is either released or absorbed in the form of heat, light, or electricity. We call the heat-releasing reactions "exothermic reactions," and the heat-absorbing reactions are referred to as "endothermic reactions." In this experiment, you are going to look at two different reactions to determine if they are exothermic or endothermic.

Procedure
1. Add 1 cup (240 mL) of hydrogen peroxide to one of the cups and 1 cup (240 mL) of warm water to the other.
2. Place a thermometer in each cup and wait 5 minutes. Record the temperature of both cups as the starting temperature in your science notebook.
3. Now, add 1 tbsp (8.5 g) of yeast to the hydrogen peroxide and gently stir. Then add 1 tbsp (14.2 g) of Epsom salts to the water and gently stir.
4. Wait until the temperature on the thermometer stabilizes and record the final temperature for the reaction in your science notebook.

Explanation
The yeast and hydrogen peroxide reaction is exothermic, and the Epsom salts and water reaction is endothermic. This is because hydrogen peroxide is a very powerful oxidizing agent,

and it immediately begins to breakdown the cell walls of the yeast. This process releases energy into the solution in the form of heat. However, it requires energy to break the bonds of the magnesium sulfate in the Epsom salts. The energy is absorbed by the magnesium and sulfate ions and removed from the water, which causes the temperature to drop.

Science Notebook
☞ Write down a short description of what you did for the experiment and add a chart with results. After that, add what you have learned from the experiment.

Online Lab - Using Energy to Observe Chemical Change

Purpose
The purpose of this online lab is to measure temperature changes in a chemical reaction and infer whether the reaction is endothermic or exothermic.

Pre-Reading
✍ Print and read the section of the workbook for the "Using Energy to Observe Chemical Change" online lab.

Procedure
✓ Do the lab entitled "Using Energy to Observe Chemical Change," and answer the questions as you work through the online lab.

Science Notebook
☞ Add the completed workbook pages that you printed to the science notebook.

Events in Science

Current Events
🕐 Find a current events article related to the field of chemistry, and complete the article summary sheet found on p. 187 of the Appendix. Once you are done, add the sheet to the events section of your science notebook.

Hands-on Activity

Optional Hands-on
✂ Use a catalyst to speed up a reaction. You will need hydrogen peroxide, dish soap, yeast, water, a bottle, and a cup. In the bottle, mix about a quarter of a cup of hydrogen peroxide with two teaspoons of dish soap. In the cup, mix a teaspoon of yeast with two tablespoons of room temperature water. Let the mixtures sit for about a minute, and observe what happens. Then add the cup mixture to the mixture in the bottle and observe what happens.
- **Explanation:** Hydrogen peroxide slowly decomposes into water and oxygen gas as soon as it is exposed to light. The yeast in this reaction acts as a catalyst to speed up this reaction, forming a lot of oxygen gas in a small amount of time. This gas is captured by the dish soap, causing the mixture to come up and out of the bottle.

Week 4 — Unit 2 (Hands-on Course) — 4-Day

Weekly Topic

→ This week will wrap up a look at chemical reactions.

	Day 1	Day 2	Day 3	Day 4
Textbook and Experiment	☐ Read *CK-12 Physical Science* Section 8.3.	☐ Read *CK-12 Physical Science* Section 8.4.	☐ Do the experiment "Exothermic or Endothermic."	☐ Do the optional hands-on science activity.
Writing	☐ Add the vocabulary to the glossary section of your science notebook.	☐ Answer the assigned questions in the reading section of your science notebook.	☐ Record what you have done in the lab section of your science notebook.	☐ Complete the Events in Science assignment.

Supplies Needed

- ✓ **Experiment** – Yeast, Hydrogen peroxide, Epsom salts, Water, 2 Cups, 2 Thermometers
- ✓ *(Optional)* Hydrogen peroxide, Dish soap, Yeast, Water, Bottle, Cup

Other Notes

Week 4 — Unit 2 (Digital Course) — 3-Day

Weekly Topic

→ This week will wrap up a look at chemical reactions.

	Day 1	Day 2	Day 3
Textbook and Experiment	☐ Read *CK-12 Physical Science* Section 8.3.	☐ Read *CK-12 Physical Science* Section 8.4.	☐ Do the online lab "Using Energy to Observe Chemical Change."
Writing	☐ Add the vocabulary to the glossary section of your science notebook.	☐ Answer the assigned questions in the reading section of your science notebook.	☐ Record what you have done in the lab section of your science notebook. ☐ Complete the Events in Science assignment.

Other Notes

Physical Science Unit 2 - Week 4

Week 5 Notes - Chemistry of Carbon, Part 1

Textbook Assignments
Reading
- *CK-12 Physical Science* Section 9.1, 9.2

Written
After you finish reading, answer questions #1, 3, 5, 7 in section 9.1 and questions #1, 3, 5, 7 in section 9.2. File your work in the reading section of your science notebook. Then define the following terms in the glossary section of your science notebook:

- ☐ Alkane
- ☐ Alkene
- ☐ Alkyne
- ☐ Isomer
- ☐ Monomer
- ☐ Polymer
- ☐ Unsaturated Hydrocarbon

Experiment - Is carbon in it?
Materials
- ✓ Sugar
- ✓ Salt
- ✓ Candle
- ✓ 2 Metal spoons
- ✓ Hot mitt

Pre-Reading
Organic compounds are substances that contain carbon. They generally have low melting and boiling points, which means that these compounds burn fairly easily in the air and always emit carbon dioxide. If there is incomplete combustion, which normally occurs in a non-laboratory environment, some elemental carbon or soot will remain. Meanwhile, inorganic compounds typically have high melting and boiling points, which means that they require a fair amount of heat energy before they will burn. In this week's experiment, you are going to use heat to test two different common household chemicals for the presence of carbon.

Procedure
1. After putting on safety goggles and protective gloves, light your candle, and let it burn for a few minutes.
   - **Caution:** Do not do this experiment without adult supervision! This experiment involves an open flame. Be sure to use the proper safety gear.
2. Meanwhile, add about ½ tsp (2.5 g) of sugar to one of the spoons. Using a pair of tongs, hold the end of the spoon with the sugar over the flame for seven minutes. Observe what happens and record what you see in your science notebook.
3. Then add about ½ tsp (2.8 g) of salt to the other spoon. Using a pair of tongs, hold the end of the spoon with the salt over the flame for seven minutes. Observe what happens and record what you see in your science notebook.

Explanation
You should see that the sugar melts, gives off a gas, and leaves a black residue. The

students should see that the salt appears to dry out, but no other change occurs.

Sugar, an organic chemical with a chemical formula of $C_6H_{12}O_6$, has a melting point of 366°F (186°C). When the heat of the candle is added, sugar readily melts, and as it continues to heat up, combustion occurs. In this process, the sugar combines with the oxygen in the air and decomposes into carbon dioxide (gas) and water (liquid), which is shown in the reaction below:

$$O_2 + C_6H_{12}O_6 \longrightarrow CO_2 + H_2O$$

In our experiment, the combustion of the sugar was incomplete which caused a black residue or soot to be left behind. This soot is composed mainly of carbon atoms, which lets us verify the presence of carbon in the sugar molecules. Salt, an inorganic chemical with a chemical formula of NaCl, has a melting point of 1474°F (801°C). So, when the heat of the candle is added, virtually no change occurs.

Science Notebook
- Write down a short description of what you did for the experiment and add the results. After that, add what you have learned from the experiment.

Online Lab - Pressure and Temperature of a Gas
Purpose
The purpose of this online lab is to explore how changing the temperature of a fixed volume of gas affects the pressure of the gas.

Pre-Reading
- Print and read the section of the workbook for the "Pressure and Temperature of a Gas" online lab.

Procedure
- Do the lab entitled "Pressure and Temperature of a Gas," and answer the questions as you work through the online lab.

Science Notebook
- Add the completed workbook pages that you printed to the science notebook.

Events in Science
Current Events
- Find a current events article related to the field of chemistry, and complete the article summary sheet found on p. 187 of the Appendix. Once you are done, add the sheet to the events section of your science notebook.

Hands-on Activity
Optional Hands-on
- Observe and compare vegetable oil and margarine. (The oil is an unsaturated compound that undergoes hydrogenation to become the saturated compound known as margarine.)

Physical Science Unit 2 - Week 5

Week 5 — Unit 2 (Hands-on Course) — 4-Day

Weekly Topic

→ This week will be an introduction to chemistry of carbon.

	Day 1	Day 2	Day 3	Day 4
Textbook and Experiment	☐ Read *CK-12 Physical Science* Section 9.1.	☐ Read *CK-12 Physical Science* Section 9.2.	☐ Do the experiment "Is carbon in it?"	☐ Do the optional hands-on science activity.
Writing	☐ Add the vocabulary to the glossary section of your science notebook.	☐ Answer the assigned questions in the reading section of your science notebook.	☐ Record what you have done in the lab section of your science notebook.	☐ Complete the Events in Science assignment.

Supplies Needed

- ✓ **Experiment** – Sugar, Salt, Candle, 2 Metal spoons, Hot mitt
- ✓ *(Optional)* Vegetable oil, Margarine

Other Notes

Physical Science Unit 2 - Week 5

Week 5	Unit 2 (Digital Course)		3-Day
Weekly Topic			
→ This week will be an introduction to chemistry of carbon.			
	Day 1	**Day 2**	**Day 3**
Textbook and Experiment	☐ Read *CK-12 Physical Science* Section 9.1.	☐ Read *CK-12 Physical Science* Section 9.2.	☐ Do the online lab "Pressure and Temperature of a Gas."
Writing	☐ Add the vocabulary to the glossary section of your science notebook.	☐ Answer the assigned questions in the reading section of your science notebook.	☐ Record what you have done in the lab section of your science notebook. ☐ Complete the Events in Science assignment.

Other Notes

Week 6 Notes - Chemistry of Carbon, Part 2

Textbook Assignments

Reading
📖 *CK-12 Physical Science* Sections 9.3, 9.4

Written
After you finish reading, answer questions #2, 4, 6 in section 9.3 and questions #1–4 in section 9.4. File your work in the reading section of your science notebook. Then define the following terms in the glossary section of your science notebook:

- ☐ Biochemical Compound
- ☐ Carbohydrate
- ☐ Cellular Respiration
- ☐ Enzyme
- ☐ Lipid
- ☐ Nucleic Acid
- ☐ Photosynthesis
- ☐ Protein

Experiment - The Enzyme Saliva

Materials
- ✓ 2 Slices of bread
- ✓ Water
- ✓ Saliva
- ✓ 2 Plastic bags

Pre-Reading
🕮 Digestion is a collection of chemical processes that occur in the body. Through these processes, the food you eat is broken down into the organic compounds that your body needs. Digestion is performed within several of the organs found in your body, and it uses different enzymes as catalysts to speed up the action along the way. In this week's experiment, you are going to look at one of your body's fluids to see if it aids in digesting your food.

Procedure
1. Label the bags #1 and #2 and then place a slice of bread into each of the bags.
2. Add ¼ cup (60 mL) of water to bag #1, seal it, and set it on the counter where it won't be disturbed.
3. Spit into a cup several times, until you have about 2 tbsp (30 mL) of saliva. Add 2 tbsp (30 mL) of water and mix well. Pour the mixture over the bread in bag #2, seal the bag, and set in on the counter next to bag #1.
4. Check the bags every 15 minutes over 4 hours and record what you see happening. Each time you check on the bread, be sure to quantify the amount of bread that has been broken down. For example, you could assign a percentage or use a scale (e.g., 1 = not

broken down, to 10 = completely broken down).
5. After 2 hours, draw conclusions and write them down in your science notebook.

Troubleshooting

If you have problems collecting enough saliva, keep your mouth open to prevent swallowing. Then think about or smell some of your favorite food, which will increase your bodies' saliva production.

Explanation

You should see that the bread with the saliva breaks down quicker and more completely than the bread with no saliva. Saliva, which is found in the mouth, serves to moisten food and protect the teeth against certain types of bacteria. This liquid contains an enzyme, called amylase, which helps to break down starch into more easily digested sugars. This is why when you chew up bread or other starch rich foods, they taste slightly sweet. Amylase is not the only enzyme used in the digestive process, there are seven enzymes altogether that aid in breaking down our food. In addition to salivary amylase, pancreatic amylase from the pancreas and maltase from the intestines also help to break down starches in our food. The digestion of proteins in our food is assisted by pepsin in the stomach, trypsin in the pancreas, and peptidase in the intestine. Finally, the process of breaking down the fats in our food is sped up by lipase from the pancreas.

Science Notebook

☞ Write down a short description of what you did for the experiment and add the results. After that, add what you have learned from the experiment.

Online Lab - Temperature and Volume of a Gas

Purpose

The purpose of this online lab is to learn about the relationship between the temperature and the volume of gas.

Pre-Reading

✎ Print and read the section of the workbook for the "Temperature and Volume of a Gas" online lab.

Procedure

✓ Do the lab entitled "Temperature and Volume of a Gas," and answer the questions as you work through the online lab.

Science Notebook

☞ Add the completed workbook pages that you printed to the science notebook.

Events in Science

Current Events
🕐 Find a current events article related to the field of chemistry, and complete the article summary sheet found on p. 187 of the Appendix. Once you are done, add the sheet to the events section of your science notebook.

Hands-on Activity

Optional Hands-on
✂ Do a starch test on several foods. You will need an iodine swab and several different types of food for testing, such as a hard-boiled egg, bread, potato, pasta, yogurt, cookies, or cheese. Cut off a sample of the food then use an iodine swab to rub some solution onto it. If the iodine changes color to blue, then starch is present. If it remains the same brown color, no starch is present. Then repeat this procedure for each of the test samples.

Week 6 — Unit 2 (Hands-on Course) — 4-Day

Weekly Topic

→ This week will wrap up a look at chemistry of carbon.

	Day 1	Day 2	Day 3	Day 4
Textbook and Experiment	☐ Read *CK-12 Physical Science* Section 9.3.	☐ Read *CK-12 Physical Science* Section 9.4.	☐ Do the experiment "The Enzyme Saliva"	☐ Do the optional hands-on science activity.
Writing	☐ Add the vocabulary to the glossary section of your science notebook.	☐ Answer the assigned questions in the reading section of your science notebook.	☐ Record what you have done in the lab section of your science notebook.	☐ Complete the Events in Science assignment.

Supplies Needed

✓ Experiment – 2 Slices of bread, Water, Saliva, 2 Plastic bags
✓ *(Optional)* An iodine swab, Several different types of food for testing, such as a hard-boiled egg, bread, potato, pasta, yogurt, cookies, or cheese

Other Notes

Week 6 — Unit 2 (Digital Course) — 3-Day

Weekly Topic

→ This week will wrap up a look at chemistry of carbon.

	Day 1	Day 2	Day 3
Textbook and Experiment	☐ Read *CK-12 Physical Science* Section 9.3.	☐ Read *CK-12 Physical Science* Section 9.4.	☐ Do the online lab "Temperature and Volume of a Gas."
Writing	☐ Add the vocabulary to the glossary section of your science notebook.	☐ Answer the assigned questions in the reading section of your science notebook.	☐ Record what you have done in the lab section of your science notebook. ☐ Complete the Events in Science assignment.

Other Notes

Physical Science Unit 2 - Week 6

Week 7 Notes - Chemistry of Solutions

Textbook Assignments

Reading
📖 *CK-12 Physical Science* Sections 10.1, 10.2, 10.3

Written
After you finish reading, answer questions #1–5 in section 10.1, questions #1–5 in section 10.2, and questions #1–4 in section 10.3. File your work in the reading section of your science notebook. Then define the following terms in the glossary section of your science notebook:

- ☐ Acidity
- ☐ Insoluble
- ☐ Neutralization reaction
- ☐ pH
- ☐ Saturated solution
- ☐ Solubility
- ☐ Soluble
- ☐ Solute
- ☐ Solvent
- ☐ Unsaturated solution

Experiment - Making Solutions

Materials
- ✓ 5 Clear cups (or beakers)
- ✓ 5 Plastic spoons
- ✓ Sugar
- ✓ Salt
- ✓ Baking soda
- ✓ Flour
- ✓ Petroleum jelly
- ✓ Water
- ✓ Vegetable oil
- ✓ Tablespoon

Pre-Reading
↪ A solution contains several different types of molecules that have been evenly mixed. Generally, a solution is made when one type of molecule, which we call the "solute," is dissolved in another, which we call the "solvent." The molecules can be either polar or nonpolar. Polar molecules have a positive and a negative end, whereas nonpolar molecules do not carry a charge. As a general rule, like dissolves like; this means that solvents with polar molecules dissolve solutes with polar molecules and vice versa. In this experiment you are going to use a polar solvent, water, and a nonpolar solvent, vegetable oil, to determine if whether several substances are polar or nonpolar.

Procedure
1. Label the cups #1 through #5. Add ½ cup (120 mL) of room temperature water to each of the five cups.

2. Then stir in 1 tbsp (12.6 g) of sugar to cup #1, 1 tbsp (17.1 g) of salt to cup #2, 1 tbsp (12 g) of baking soda to cup #3, 1 tbsp (5 g) of flour to cup #4, and 1 tbsp (8 g) of petroleum jelly to cup #5.
3. Gently stir each of the cups with a spoon, using a different spoon for each cup. Wait 10 minutes and record your observations and results in your science notebook.
4. Pour out the solutions, rinse well and dry each of the cups.
5. Add ½ cup (120 mL) of room temperature oil to each of the five cups. Repeat steps 2 and 3, mixing the test solutes with the oil solvent instead.

Explanation

You should find that salt, baking soda, and sugar are polar molecules and that flour and petroleum jelly are both nonpolar molecules. Salt and baking soda have ionic bonds, meaning the atoms have exchanged electrons to form a chemical bond that holds the molecule together through the associated charges. These two molecules are highly polar substances that can easily dissolve in a polar solvent. Sugar has a long carbon chain that has no associated charge and an OH group that carries a charge. However, it is polar because the pull of the OH group is stronger than the carbon chain, so sugar dissolves better in water than in oil. Flour is composed of thousands of different molecules that come from grinding up grain. Some of these molecules are polar and some are not. On the whole, cooking flour contains more nonpolar molecules, so it distributes more evenly in the oil. Petroleum jelly is a mixture of hydrocarbons, all of which are nonpolar, which is why it dissolves easily in the oil and not in the water.

Science Notebook

☞ Write down a short description of what you did for the experiment, and add a chart for the results. After that, add what you have learned from the experiment.

Online Lab - Molecular Motion and Thermal Energy

Purpose

The purpose of this online lab is to calculate the amount of kinetic energy in a system and observe what happens when the system is given more energy.

Pre-Reading

☞ Print and read the section of the workbook for the "Molecular Motion and Thermal Energy" online lab.

Procedure

✓ Do the lab entitled "Molecular Motion and Thermal Energy," and answer the questions as you work through the online lab.

Science Notebook

☞ Add the completed workbook pages that you printed to the science notebook.

Events in Science

Current Events
🕐 Find a current events article related to the field of chemistry, and complete the article summary sheet found on p. 187 of the Appendix. Once you are done, add the sheet to the events section of your science notebook.

Hands-on Activity

Optional Hands-on
✂ Do a kitchen acid/base test using the directions from the following blog post:
💻 https://elementalscience.com/blogs/science-activities/kitchen-acid-test
You will need a head of red cabbage, water, and a pot, plus several cups and a variety of liquids or powders from your kitchen to test, such as lemon juice, baking soda, soda, and detergent.

Week 7 — Unit 2 (Hands-on Course) — 4-Day

Weekly Topic

→ This week will look at the chemistry of solutions.

	Day 1	Day 2	Day 3	Day 4
Textbook and Experiment	☐ Read *CK-12 Physical Science* Section 10.1 and 10.2.	☐ Read *CK-12 Physical Science* Section 10.3.	☐ Do the experiment "Making Solutions"	☐ Do the optional hands-on science activity.
Writing	☐ Add the vocabulary to the glossary section of your science notebook.	☐ Answer the assigned questions in the reading section of your science notebook.	☐ Record what you have done in the lab section of your science notebook.	☐ Complete the Events in Science assignment.

Supplies Needed

- ✓ **Experiment** – 5 Clear cups (or beakers), 5 Plastic spoons, Sugar, Salt, Baking soda, Flour, Petroleum jelly, Water, Vegetable oil, Tablespoon
- ✓ *(Optional)* A head of red cabbage, Water, Pot, A variety of liquids or powders from your kitchen to test, such as lemon juice, baking soda, soda, or detergent, Several cups

Other Notes

Week 7	Unit 2 (Digital Course)		3-Day

Weekly Topic

→ This week will look at the chemistry of solutions.

	Day 1	Day 2	Day 3
Textbook and Experiment	☐ Read *CK-12 Physical Science* Section 10.1 and 10.2.	☐ Read *CK-12 Physical Science* Section 10.3.	☐ Do the online lab "Molecular Motion and Thermal Energy."
Writing	☐ Add the vocabulary to the glossary section of your science notebook.	☐ Answer the assigned questions in the reading section of your science notebook.	☐ Complete the Events in Science assignment.

Other Notes

Week 8 Notes - Nuclear Chemistry

Textbook Assignments

Reading
📖 *CK-12 Physical Science* Sections 11.1, 11.2, 11.3

Written
After you finish reading, answer questions #1–6 in section 11.1, questions #1–3, 5 in section 11.2, and questions #1, 3, 5, 7 in section 11.3. File your work in the reading section of your science notebook. Then define the following terms in the glossary section of your science notebook:

- ☐ Half-life
- ☐ Nuclear energy
- ☐ Nuclear fission
- ☐ Nuclear fusion
- ☐ Radiation
- ☐ Radioactive dating
- ☐ Radioactive decay
- ☐ Radioactivity
- ☐ Radioisotope

Unit Test
Take Unit Test #3 – Chemical Interactions starting on *CK-12 Physical Science*, p. 343.

Experiment
☞ There is no hands-on experiment scheduled for this week.

Online Lab
☞ There is no online lab scheduled for this week.

Events in Science

Current Events
🕒 Find a current events article related to the field of chemistry, and complete the article summary sheet found on p. 187 of the Appendix. Once you are done, add the sheet to the events section of your science notebook.

Hands-on Activity

Optional Hands-on
✂ Do the "Alpha Decay" simulation from PhET. Here is the link to the online demonstration:
  💻 http://phet.colorado.edu/en/simulation/alpha-decay
✂ Do the "Nuclear Fission" simulation from PhET. Here is the link to the online demonstration:
  💻 http://phet.colorado.edu/en/simulation/nuclear-fission

Week 8 — Unit 2 (Hands-on Course) — 4-Day

Weekly Topic

→ This week will look at nuclear chemistry.

	Day 1	Day 2	Day 3	Day 4
Textbook and Experiment	☐ Read *CK-12 Physical Science* Section 11.1.	☐ Read *CK-12 Physical Science* Section 11.2.	☐ Read *CK-12 Physical Science* Section 11.3.	☐ Do the optional hands-on science activity. ☐ Take Unit Test #3 – Chemical Interactions starting on *CK-12 Physical Science*, p. 343.
Writing	☐ Add the vocabulary to the glossary section of your science notebook.	☐ Answer the assigned questions in the reading section of your science notebook.	☐ Answer the assigned questions in the reading section of your science notebook.	☐ Complete the Events in Science assignment.

Supplies Needed

Other Notes

Week 8 — Unit 2 (Digital Course) — 3-Day

Weekly Topic

→ This week will look at nuclear chemistry.

	Day 1	Day 2	Day 3
Textbook and Experiment	☐ Read *CK-12 Physical Science* Section 11.1 and 11.2.	☐ Read *CK-12 Physical Science* Section 11.3.	☐ *(Optional)* Do the hands-on science activity. ☐ Take Unit Test #3 – Chemical Interactions starting on *CK-12 Physical Science*, p. 343.
Writing	☐ Add the vocabulary to the glossary section of your science notebook.	☐ Answer the assigned questions in the reading section of your science notebook.	☐ Complete the Events in Science assignment.

Other Notes

Unit 2 - Answers

Week 1

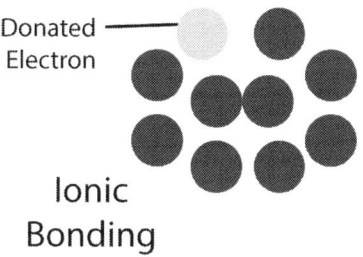

Ionic Bonding

Experiment
- Your results should look like this:

Science Notebook Assignment (Experiment)
- MgO – magnesium oxide
- NaBr – sodium bromide
- Li2S – lithium sulfide
- $MgSO_4$ – magnesium sulfate
- $Be(OH)_2$ – beryllium hydroxide
- $Sr(NO_3)_2$ – strontium nitrate

Week 2

Experiment
- Your results should look like this:

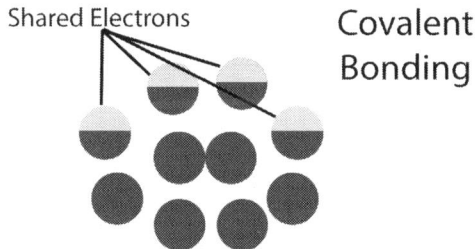

Covalent Bonding

Science Notebook Assignment (Experiment)
- CO_2 – carbon dioxide
- NO_2 – nitrogen dioxide

- SO_3 – sulfur trioxide
- N_2S – dinitrogen sulfide
- BF_3 – boron trifluoride
- P_2Br_4 – diphosphorus tetrabromide

Week 3

Science Notebook Assignment (Experiment)

1. $2\ AgNO_3 + Cu \longrightarrow Cu(NO_3)_2 + 2\ Ag$
2. $AlBr_3 + 3\ K \longrightarrow 3\ KBr + Al$
3. $2\ LiNO_3 + CaBr_2 \longrightarrow Ca(NO_3)_2 + 2\ LiBr$
4. $PbBr_2 + 2\ HCl \longrightarrow 2\ HBr + PbCl_2$
5. $2\ NaCN + CuCO_3 \longrightarrow Na_2CO_3 + Cu(CN)_2$
6. $2\ Al + 6\ HCl \longrightarrow 3\ H_2 + 2\ AlCl_3$
7. $N_2 + 2\ O_2 \rightleftharpoons 2\ NO_2$
8. $2\ NO_2 \rightleftharpoons N_2O_4$
9. $2\ SO_3 \rightleftharpoons 2\ SO_2 + O_2$
10. $N_2 + 3\ H_2 \rightleftharpoons 2\ NH_3$

Week 7

Experiment

- Your results chart should look like this:

Solute	Solvent: Water	Solvent: Vegetable Oil
Sugar	SS	NS
Salt	S	NS
Baking soda	S	NS
Flour	NS	SS
Petroleum jelly	NS	S

*S = Soluble, SS = Semi-soluble, NS = Not soluble

Physical Science

Unit 3 – Forces and Motion

Unit 3 - Overview of Study

Topics Covered

- **Week 1:** Motion
- **Week 2:** Forces, Part 1
- **Week 3:** Forces, Part 2
- **Week 4:** Newton's Laws of Motion
- **Week 5:** Fluid Forces
- **Week 6:** Work and Machines, Part 1
- **Week 7:** Work and Machines, Part 2

Supplies Needed (for the Hands-on Course)

Week	Experiment	Optional Activity
1	Cardboard or plastic track, Blocks or thick books, Toy car, Stopwatch	No supplies needed
2	Rubber band, Small wooden block (aka. Jenga block), Eye-hook screw, Sandpaper, Felt, Foil, Spray oil, Tape measure	Cardboard or plastic track, Sand or salt, Marbles, Stopwatch
3	A raw egg, Various shock absorbing materials (such as cotton balls, newspaper, packing peanuts or fabric), 1 Quart plastic container (the type that fruit is typically packed in), Masking tape	Object from nature
4	Jenga block with the eyehook from week 2, String, 2 Toy cars, Egg	Ropes, Several friends
5	Balloon, Old CD, 2–L Soda bottle lid, Thin nail, Glue	Cup, Water, Egg, Salt
6	Several LEGO® bricks, including wheels and axles, 3 Balloons	No supplies needed
7	Materials will vary based on what you choose to build	No supplies needed

Week 1 Notes - Motion

Textbook Assignments
Reading
📖 *CK-12 Physical Science* Sections 12.1, 12.2, 12.3

Written
After you finish reading, answer questions #1–4 in section 12.1, questions #1, 3, 5 in section 12.2, and questions #1–3, 5 in section 12.3. File your work in the reading section of your science notebook. Then define the following terms in the glossary section of your science notebook:

- ☐ Acceleration
- ☐ Distance
- ☐ Frame reference
- ☐ Motion
- ☐ Speed
- ☐ Vector
- ☐ Velocity

Experiment - Does height affect speed?
Materials
- ✓ Cardboard or plastic track
- ✓ Blocks or thick books
- ✓ Toy car
- ✓ Stopwatch

Pre-Reading
We know from Newton's laws of motion that a force must act on a vehicle to get it moving. We also know that the greater the force, the quicker the vehicle will go. We use speed to measure how fast that vehicle is traveling. To calculate the average speed of a car, we need the distance it traveled and the time it took. The equation looks like this:

$$\text{Average Speed } (v) = \text{total distance } (d) \, / \, \text{total time } (t)$$

In today's experiment, you are going to measure the time it takes for a toy car to go down a given length of track. You will also vary the height of the track to see how the force of gravity affects the speed of the car.

Procedure
1. Use the cardboard (or plastic track) to build a track that is 1 meter long. Use the blocks (or books) to prop the track up so that one end of it is 15 centimeters higher than the other.
2. Now, use the stopwatch to measure the time it takes for the car to go from the top of the track to the end. Record the time on the science notebook and then repeat two more times for a total of three trials at 15 centimeters.
3. Next, add several more blocks (or books) to prop the track up so that one end of it is 30 centimeters higher than the other.
4. Use the stopwatch to measure the time it takes for the car to go from the top of the track to the end. Record the time on the science notebook and then repeat two more times for a total of three trials at 30 centimeters.

5. Finally, add a few more blocks (or books) to prop the track up so that one end of it is 45 centimeters higher than the other.
6. Use the stopwatch to measure the time it takes for the car to go from the top of the track to the end. Record the time on the science notebook and then repeat two more times for a total of three trials at 45 centimeters.
7. Draw conclusions and add them to the science notebook.

Troubleshooting Tips
Make sure that the track is smooth, even, and free from debris that might slow down the progress of the car.

Explanation
For the car to move down the track a force must either push or pull. In this experiment, the force of gravity overcomes the forces of friction and inertia to send the car heading down the track. As the height of the track increases, the force of gravity that pulls on the car also increases. This causes the speed of the toy car to increase as the height of the track increases.

Science Notebook
☞ Write down a short description of what you did for the experiment and add a chart with the results. Calculate the average speed of the car for each trial. Here is a sample calculation of average speed for the following chart:

	Trial #1	Trial #2	Trial #3
Track at Height of 30 cm	5 seconds	6 seconds	5 seconds

1. Calculate the average time (t) = (5 + 6 + 5) / 3 = 5.33 s
2. Calculate the height in meters (d) = 30 cm / 100 cm per m = 0.3 m
3. Calculate the average speed = d / t = (0.3) / (5.33) = 0.056 m/s

After that, add what you have learned from the experiment.

Online Lab - Graphing Motion
Purpose
The purpose of this online lab is to learn how to use graphs to describe the motion of objects.

Pre-Reading
☞ Print and read the section of the workbook for the "Graphing Motion" online lab.

Procedure
✓ Do the lab entitled "Graphing Motion," and answer the questions as you work through the online lab.

Science Notebook
☞ Add the completed workbook pages that you printed to the science notebook.

Events in Science
Current Events
🕐 Find a current events article related to the field of physics, and complete the article summary sheet found on p. 187 of the Appendix. Once you are done, add the sheet to the events section of your science notebook.

Hands-on Activity
Optional Hands-on
✂ There is no optional hands-on activity for this week.

Week 1	Unit 3 (Hands-on Course)			4-Day
Weekly Topic				
→ This week will look at motion.				
	Day 1	Day 2	Day 3	Day 4
Textbook and Experiment	☐ Read *CK-12 Physical Science* Section 12.1 and 12.2.	☐ Read *CK-12 Physical Science* Section 12.3.	☐ Do the experiment "Does height affect speed?"	
Writing	☐ Add the vocabulary to the glossary section of your science notebook.	☐ Answer the assigned questions in the reading section of your science notebook.	☐ Record what you have done in the lab section of your science notebook.	☐ Complete the Events in Science assignment.

Supplies Needed

✓ **Experiment** – Cardboard or plastic track, Blocks or thick books, Toy car, Stopwatch

Other Notes

Physical Science Unit 3 - Week 1

Week 1		Unit 3 (Digital Course)		3-Day
Weekly Topic				
→ This week will look at motion.				
	Day 1		**Day 2**	**Day 3**
Textbook and Experiment	☐ Read *CK-12 Physical Science* Section 12.1 and 12.2.		☐ Read *CK-12 Physical Science* Section 12.3.	☐ Do the online lab "Graphing Motion."
Writing	☐ Add the vocabulary to the glossary section of your science notebook.		☐ Answer the assigned questions in the reading section of your science notebook.	☐ Record what you have done in the lab section of your science notebook. ☐ Complete the Events in Science assignment.
Other Notes				

Week 2 Notes - Forces, Part 1

Textbook Assignments
Reading
📖 *CK-12 Physical Science* Sections 13.1, 13.2

Written
After you finish reading, answer questions #1–5 in section 13.1 and questions #2, 4, 6 in section 13.2. File your work in the reading section of your science notebook. Then define the following terms in the glossary section of your science notebook:

- ☐ Fluid
- ☐ Force
- ☐ Friction
- ☐ Net force
- ☐ Newton

Experiment - Friction
Materials
- ✓ Rubber band
- ✓ Small wooden block (i.e., Jenga block)
- ✓ Eye-hook screw
- ✓ Sandpaper
- ✓ Felt
- ✓ Foil
- ✓ Spray oil
- ✓ Tape measure

Pre-Reading
☞ When an object is in forward motion, several forces are acting on it. There is the driving force that is propelling the object forward. There is weight (or gravity) that pulls the object downward. There is air resistance that slows the object down. Finally, there is friction. In today's experiment, you are going to act as the driving force for a block as it moves across a track. Then you are going to use a variety of materials to test how friction affects the motion of the block.

Procedure
1. Screw the eye-hook screw into the top of the wooden block. Then attach it to the hook on the rubber band so that the block can be dragged horizontally.
2. Use the tape measure to mark off a 1 foot (0.3 meter) track on a smooth surface such as a table our counter.
3. Now, place the block at the beginning of your track with the force meter in front over the track. Pull the block using the rubber band evenly to the end in three seconds. Observe how much the rubber band is stretched and record that on your science notebook.
4. Then place the piece of sandpaper on your track. Like before, put the block at the beginning of the track and pull it evenly to the end in three seconds. Observe how much the rubber band is stretched and record that on your science notebook.
5. Repeat with the same process using the felt, putting the block at the beginning of the track, and pulling it evenly to the end in three seconds. Observe how much the rubber

band is stretched and record that on your science notebook.
6. Finally, place the foil over the track and coat it well with spray oil. Then, as before, put block at the beginning of your track and pull it evenly to the end in three seconds. Observe how much the rubber band is stretched and record that on your science notebook.
7. Draw conclusions and add them to the science notebook.

Explanation

You should see that more force was needed to pull the block when it was on the felt and sandpaper. You should also that less force was needed to pull the block when it was on the oil-covered foil. Both the felt and the sandpaper increase the amount of friction that acts on the block as it slides over the track. The oil-coated foil reduces the amount of friction that acts on the block as it slides over the track. Friction is a force that opposes the motion of an object as it passes another. So, when friction increases, the object will slow down. Conversely, when friction is decreased, the object will speed up.

Science Notebook

- Write down a short description of what you did for the experiment and add a chart with results. After that, add what you have learned from the experiment.
- Complete the resultant force worksheet practice in the Appendix on p. 197. Add your work and answers to the science notebook.

Online Lab - Balanced Forces

Purpose

The purpose of this online lab is to understand the effects of balanced and unbalanced forces on an object.

Pre-Reading

Print and read the section of the workbook for the "Balanced Forces" online lab.

Procedure

✓ Do the lab entitled "Balanced Forces," and answer the questions as you work through the online lab.

Science Notebook

- Add the completed workbook pages that you printed to the science notebook.

Events in Science

Current Events

Find a current events article related to the field of physics, and complete the article summary sheet found on p. 187 of the Appendix. Once you are done, add the sheet to the events section of your science notebook.

Physical Science Unit 3 - Week 2

Hands-on Activity
Optional Hands-on

✂ Study the effect of friction using a marble track, just like Galileo did. You will need a cardboard or plastic track, sand or salt, marbles, and a stopwatch Set up a track similar to the one in the experiment and send the marble down it several times. Each time record the time it takes to get to the bottom. Then sprinkle some sand or salt all over the track. Send the marbles down several more times, recording the time it takes.

- **Explanation:** You should see that the marbles were much slower the second time because of the amount of friction that was created by the sand or salt on the track.

Week 2 — Unit 3 (Hands-on Course) — 4-Day

Weekly Topic

→ This week will be an introduction to forces.

	Day 1	Day 2	Day 3	Day 4
Textbook and Experiment	☐ Read *CK-12 Physical Science* Section 13.1.	☐ Read *CK-12 Physical Science* Section 13.2.	☐ Do the experiment "Friction."	☐ Do the optional hands-on science activity.
Writing	☐ Add the vocabulary to the glossary section of your science notebook.	☐ Answer the assigned questions in the reading section of your science notebook.	☐ Record what you have done in the lab section of your science notebook.	☐ Complete the Events in Science assignment.

Supplies Needed

✓ **Experiment** – Rubber band, Small wooden block (aka. Jenga block), Eye-hook screw, Sandpaper, Felt, Foil, Spray oil, Tape measure
✓ *(Optional)* Cardboard or plastic track, Sand or salt, Marbles, Stopwatch

Other Notes

Week 2 — Unit 3 (Digital Course) — 3-Day

Weekly Topic

→ This week will be an introduction to forces.

	Day 1	Day 2	Day 3
Textbook and Experiment	☐ Read *CK-12 Physical Science* Section 13.1.	☐ Read *CK-12 Physical Science* Section 13.2.	☐ Do the online lab "Balanced Forces." ☐ *(Optional)* Do the hands-on science activity.
Writing	☐ Add the vocabulary to the glossary section of your science notebook.	☐ Answer the assigned questions in the reading section of your science notebook.	☐ Record what you have done in the lab section of your science notebook. ☐ Complete the Events in Science assignment.

Other Notes

Week 3 Notes - Forces, Part 2

Textbook Assignments
Reading
📖 *CK-12 Physical Science* Section 13.3, 13.4

Written
After you finish reading, answer questions #1, 3, 5 in section 13.3 and questions #1–4 in section 13.4. File your work in the reading section of your science notebook. Then define the following terms in the glossary section of your science notebook:

- ☐ Elastic force
- ☐ Elasticity
- ☐ Gravity
- ☐ Law of universal gravity
- ☐ Orbit
- ☐ Projectile motion

Experiment - Egg-drop
Materials
- ✓ A raw egg
- ✓ Various shock absorbing materials (such as cotton balls, newspaper, packing peanuts or fabric)
- ✓ 1 Quart plastic container (the type that fruit is typically packed in)
- ✓ Masking tape

Pre-Reading
Inertia and momentum are often confused or used interchangeably. However, the two are quite different. The amount of inertia force an object experiences is only based on the object's mass, whereas the force of momentum an object feels is dependent upon its mass and its speed. In other words, inertia is how much something resists changes in motion while momentum increases or decreases with motion. In this experiment, you will test how to slow down the inertia and momentum of a falling egg so that the "reaction" from the third law of motion doesn't result in a cracked egg.

Procedure
1. Begin by examining the different shock absorbing materials you have and using them to fill the 1 quart container in such a way that the egg will be protected as it falls. Be sure to have the egg on hand so that you can measure it and make sure the egg fits in the remaining space.
2. Tape across the top of the container to hold in the egg and the materials.
3. Now, hold your containers over your head and drop it. Observe what happens. Did the egg crack?
4. Draw conclusions and add them to the science notebook.

Explanation
As the egg drops, its speed increases, which causes the force of momentum to increase. When it hits the ground, an equal shock force (Newton's 3rd Law of Motion) is sent back

into the egg. If the egg was not protected, it would surely crack, as it did in the experiment from earlier this week. However, in this case, the materials surrounding the egg absorb most of the shock force, so that the force that the egg eventually felt was minimal.

Science Notebook
☞ Write down a short description of what you did for the experiment and add the results. After that, add what you have learned from the experiment.

Online Lab
☞ There is no online lab scheduled for this week.

Events in Science
Current Events
🕒 Find a current events article related to the field of physics, and complete the article summary sheet found on p. 187 of the Appendix. Once you are done, add the sheet to the events section of your science notebook.

Hands-on Activity
Optional Hands-on
✂ Test gravity using several objects from nature. Begin by taking a walk outside to look for several objects in nature that are round and about the same size. The objects should have different weights, such as a piece of fruit, a rock, and a nut. Once you get home, hold each of the round objects in your hands and drop them at the same time. What happened?
- **Explanation:** You should see that both of the objects hit the ground at the very same time. If you can do this safely from a porch or balcony that will give you a bit more height, your results will be even more amazing.

Week 3 — Unit 3 (Hands-on Course) — 4-Day

Weekly Topic

→ This week will continue the introduction to forces.

	Day 1	Day 2	Day 3	Day 4
Textbook and Experiment	☐ Read *CK-12 Physical Science* Section 13.3.	☐ Read *CK-12 Physical Science* Section 13.4.	☐ Do the experiment "Egg-drop."	☐ Do the optional hands-on science activity.
Writing	☐ Add the vocabulary to the glossary section of your science notebook.	☐ Answer the assigned questions in the reading section of your science notebook.	☐ Record what you have done in the lab section of your science notebook.	☐ Complete the Events in Science assignment.

Supplies Needed

✓ **Experiment** – A raw egg, Various shock absorbing materials (such as cotton balls, newspaper, packing peanuts or fabric), 1 Quart plastic container (the type that fruit is typically packed in), Masking tape
✓ **(Optional)** Objects from nature

Other Notes

Week 3	Unit 3 (Digital Course)		3-Day
Weekly Topic			
→ This week will continue the introduction to forces.			
	Day 1	Day 2	Day 3
Textbook and Experiment	☐ Read *CK-12 Physical Science* Section 13.3.	☐ Read *CK-12 Physical Science* Section 13.4.	☐ *(Optional)* Do the hands-on science activity.
Writing	☐ Add the vocabulary to the glossary section of your science notebook.	☐ Answer the assigned questions in the reading section of your science notebook.	☐ Complete the Events in Science assignment.
Other Notes			

Week 4 Notes - Newton's Law of Motion

Textbook Assignments
Reading
- *CK-12 Physical Science* Sections 14.1, 14.2, 14.3

Written
After you finish reading, answer questions #1–4 in section 14.1, questions #1–5 in section 14.2, and questions #2, 4, 6 in section 14.3. File your work in the reading section of your science notebook. Then define the following terms in the glossary section of your science notebook:

- ☐ Inertia
- ☐ Law of conservation of momentum
- ☐ Momentum
- ☐ Newton's first law of motion
- ☐ Newton's second law of motion
- ☐ Newton's third law of motion

Experiment - The Three Laws
Materials
- ✓ Jenga block with the eyehook from week 2
- ✓ String
- ✓ 2 Toy cars
- ✓ Egg

Pre-Reading
Isaac Newton built on Galileo's work on friction and motion through a number of experiments. These tests led to his development of the three laws of motion. The laws state:

1. An object will not move unless a force like a push or pull moves it. Once it is moving, an object will not stop moving in a straight line unless it's forced to change.
2. The greater the force on an object, the greater the change in its motion. The greater the mass of an object, the greater the force needed to change its motion.
3. For every action, there is an equal but opposite reaction.

In today's experiment, you are going to do three tests where you will see each of the laws in action.

Procedure
1. Test the three laws of motion.
   - Test # 1 – You will need a Jenga block with the eyehook and string. Tie the string to the block and place it on a smooth surface. Pull the block along with a decent amount of force and then stop suddenly. Observe what happens to the block.
   - Test # 2 – You will need the two toy cars and a partner for this test. Line up the two cars evenly on a flat surface. Gently push your car forward while your partner pushes his car forward with a greater force at the same time. Observe what happens to the two cars.
   - Test # 3 – You will need an egg for this test. Head outside with the egg. Drop the egg

onto the pavement from a height of four to five feet and observe what happens.
2. Draw conclusions and add them to the science notebook.

Explanation
For test one, you should see that when they stopped suddenly, the block continued to move until the string went taut. Once the string was tight, the block bounced back a bit and eventually stopped. For test two, you should see that the car that was pushed with greater force moved faster and farther than the other car. For test three, you should see that the egg cracked and splattered on the pavement.

In test one, the students are looking at inertia from the first law of motion. The block began to move because the student pulled the string attached to it. When you stop suddenly, the block continues to move because the force of the sudden stop in the string has not acted on it yet. However, when then string goes taut, the force is strong enough to change the motion of the block and eventually stop it. In test two, you should see that the force that acts on the second car is greater than the force that acts on the first one. This causes a greater change in the second car's motion, which means it goes faster and farther. In test three, the egg reaches its terminal velocity before it hits the pavement. It is moving with such a force that the action of the sudden stop causes an equally violent reaction. This reaction has the force to break apart the egg and splatter it on the pavement.

Science Notebook
- Write down a short description of what you did for the experiment and add the results. Finally, add what you have learned from the experiment.
- Complete the second law of motion worksheet practice in the Appendix on p. 198. Add your work and answers to the science notebook.

Online Lab - Acceleration and Friction
Purpose
The purpose of this online lab is to investigate the effect that friction has on the acceleration of an object.
Pre-Reading
- Print and read the section of the workbook for the "Acceleration and Friction" online lab.
Procedure
- Do the lab entitled "Acceleration and Friction," and answer the questions as you work through the online lab.
Science Notebook
- Add the completed workbook pages that you printed to the science notebook.

Events in Science
Current Events
- Find a current events article related to the field of physics, and complete the article summary sheet found on p. 187 of the Appendix. Once you are done, add the sheet to the

events section of your science notebook.

Hands-on Activity
Optional Hands-on

- In a tug of war, each team is using force to pull the other team across the line. One team's pulling force cancels out the other team's pulling force, which keeps the players at a stand-still. That is, until one team's pulling force is greater than the other's! This week, observe how force plays a role in tug of war as you try play it. If you can't get a team together, do the tug of war simulation from the PhET website.
  - https://phet.colorado.edu/en/simulation/forces-and-motion-basics

Week 4	Unit 3 (Hands-on Course)			4-Day
Weekly Topic				
→ This week will begin a look at Newton's Laws of Motion.				
	Day 1	**Day 2**	**Day 3**	**Day 4**
Textbook and Experiment	❏ Read *CK-12 Physical Science* Section 14.1 and 14.2.	❏ Read *CK-12 Physical Science* Section 14.3.	❏ Do the experiment "The Three Laws."	❏ Do the optional hands-on science activity.
Writing	❏ Add the vocabulary to the glossary section of your science notebook.	❏ Answer the assigned questions in the reading section of your science notebook.	❏ Record what you have done in the lab section of your science notebook.	❏ Complete the Events in Science assignment.
Supplies Needed				
✓ **Experiment** – Jenga block with the eyehook from week 2, String, 2 Toy cars, Egg ✓ *(Optional)* Rope, Several friends				
Other Notes				

Physical Science Unit 3 - Week 4

Week 4 — Unit 3 (Digital Course) — 3-Day

Weekly Topic

→ This week will begin a look at Newton's Laws of Motion.

	Day 1	Day 2	Day 3
Textbook and Experiment	☐ Read *CK-12 Physical Science* Section 14.1 and 14.2.	☐ Read *CK-12 Physical Science* Section 14.3.	☐ Do the online lab "Acceleration and Friction." ☐ *(Optional)* Do the hands-on science activity.
Writing	☐ Add the vocabulary to the glossary section of your science notebook.	☐ Answer the assigned questions in the reading section of your science notebook.	☐ Record what you have done in the lab section of your science notebook. ☐ Complete the Events in Science assignment.

Other Notes

Week 5 Notes - Fluid Forces

Textbook Assignments
Reading
📖 *CK-12 Physical Science* Section 15.1, 15.2

Written
After you finish reading, answer questions #2, 4, 6 in section 15.1 and questions #1, 3, 5 in section 15.2. File your work in the reading section of your science notebook. Then define the following terms in the glossary section of your science notebook:

- ☐ Archimedes' law
- ☐ Bernoulli's law
- ☐ Buoyancy
- ☐ Buoyant force
- ☐ Pascal
- ☐ Pascal's law

Experiment - Hovercraft
Materials
- ✓ Balloon
- ✓ Old CD
- ✓ 2–L Soda bottle lid
- ✓ Thin nail
- ✓ Glue

Pre-Reading
Liquids and gases are both classified as fluids. As fluids, they exert pressure on the containers they are in. They can also flow from place to place, exerting a force as they go. In this experiment, you are going to use the force of a fluid to move a hovercraft.

Procedure
1. Begin by using the nail to poke at least 10 holes in the top of your soda bottle lid.
2. Then use the glue to attach the bottom of the lid over the hole in the center of the CD so that the holes face up.
3. After the glue dries, set the CD on a table or flat surface and blow up the balloon.
4. Quickly fit the balloon over the soda bottle lid, and watch what happens!
5. Draw conclusions and add them to the science notebook.

Explanation
The CD will rise up just a bit and move across the table, just like a hovercraft. When the air is expelled from the balloon, it creates a cushion between the surface and the CD. This makes it possible for the CD-ship to float or hover over the surface. You can also repeat this experiment using a hair dryer on low instead so that you can direct the movement of the CD-ship. If you have time, test out a different size of craft by using a paper plate.

Science Notebook
☞ Write down a short description of what you did for the experiment and add the results. After that, add what you have learned from the experiment.

Online Lab - Friction and Motion
Purpose
The purpose of this online lab is to discover how friction affects the motion of a sliding object.
Pre-Reading
- Print and read the section of the workbook for the "Friction and Motion" online lab.
Procedure
- Do the lab entitled "Friction and Motion," and answer the questions as you work through the online lab.
Science Notebook
- Add the completed workbook pages that you printed to the science notebook.

Events in Science
Current Events
- Find a current events article related to the field of physics, and complete the article summary sheet found on p. 187 of the Appendix. Once you are done, add the sheet to the events section of your science notebook.

Hands-on Activity
Optional Hands-on
- Float an egg. You will need a cup, water, egg, and salt. Place an egg in a cup filled about two-thirds of the way with tap water and observe what happens. (It should sink to the bottom.) Add about a quarter to a half a cup of salt, depending on the size of your glass. Stir gently and observe what happens to the egg.
  - **Explanation:** You should see the egg float. If it does not, you may need to add a bit more salt. The egg floats in the salt water because the liquid is more dense, which means that the egg weighs less than the salt water it displaces.

Week 5	Unit 3 (Hands-on Course)			4-Day
Weekly Topic				

→ This week will look at forces in fluids.

	Day 1	Day 2	Day 3	Day 4
Textbook and Experiment	☐ Read *CK-12 Physical Science* Section 15.1.	☐ Read *CK-12 Physical Science* Section 15.2.	☐ Do the experiment "Hovercraft."	☐ Do the optional hands-on science activity.
Writing	☐ Add the vocabulary to the glossary section of your science notebook.	☐ Answer the assigned questions in the reading section of your science notebook.	☐ Record what you have done in the lab section of your science notebook.	☐ Complete the Events in Science assignment.

Supplies Needed

- ✓ **Experiment** – Balloon, Old CD, 2-L Soda bottle lid, Thin nail, Glue
- ✓ *(Optional)* Cup, Water, Egg, Salt

Other Notes

Week 5	Unit 3 (Digital Course)	3-Day

Weekly Topic

→ This week will look at forces in fluids.

	Day 1	Day 2	Day 3
Textbook and Experiment	❒ Read *CK-12 Physical Science* Section 15.1.	❒ Read *CK-12 Physical Science* Section 15.2.	❒ Do the online lab "Friction and Motion." ❒ *(Optional)* Do the hands-on science activity.
Writing	❒ Add the vocabulary to the glossary section of your science notebook.	❒ Answer the assigned questions in the reading section of your science notebook.	❒ Record what you have done in the lab section of your science notebook. ❒ Complete the Events in Science assignment.

Other Notes

Week 6 Notes - Work and Machines, Part 1

Textbook Assignments
Reading
 📖 *CK-12 Physical Science* Sections 16.1, 16.2
Written
 After you finish reading, answer questions #1–5 in section 16.1 and questions #1, 3, 5, 7 in section 16.2. File your work in the reading section of your science notebook. Then define the following terms in the glossary section of your science notebook:
- ☐ Efficiency
- ☐ Joule
- ☐ Machine
- ☐ Mechanical advantage
- ☐ Power
- ☐ Watt
- ☐ Work

Experiment - LEGO Bricks Balloon Car
Materials
- ✓ Several LEGO bricks, including wheels and axles
- ✓ 3 Balloons

Pre-Reading
 ✎ The LEGO bricks balloon car is basically a simple machine that uses wind power to move. The wheels and axles act as simple machines. They reduce the amount a friction the vehicle experiences as it travels across a surface, which makes it easier to move. But Newton's First Law of Motion states that an object will not move unless a force acts upon it. So, we have to have a force that pushes the car to get it to move. In this experiment, you will make three LEGO bricks balloon cars and test them.

Procedure
1. Begin by building three cars, all with space to add a balloon to power them. Follow these three rules when building the cars:
   - The car must be vertically and horizontally stable.
   - The balloon needs to have enough space to inflate.
   - The wheels should be under the car so the balloon doesn't hit them.
2. Insert the balloons into the space you left, line up the cars, blow up the balloons, and let them go.
3. Draw conclusions and add them to the science notebook.

Explanation
 In the case of the LEGO bricks balloon car, we are using a power source, the balloon! When you blow up a balloon the elastic stretches, which creates a globe full of pressurized air. If you don't tie off the end, the stretched elastic quickly forces the air inside out the tiny hole at the end of the balloon. So, in the LEGO bricks balloon car, the air escapes causing an equal but opposite reaction that pushes the car forward. The less our car weighs, the less friction or drag it creates, and the faster it goes.

Science Notebook
☞ Write down a short description of what you did for the experiment and add the results. After that, add what you have learned from the experiment.

Online Lab - Energy of Combustion
Purpose
The purpose of this online lab is to discover how energy changes during a chemical reaction.
Pre-Reading
᧞ Print and read the section of the workbook for the "Energy of Combustion" online lab.
Procedure
✓ Do the lab entitled "Energy of Combustion," and answer the questions as you work through the online lab.
Science Notebook
☞ Add the completed workbook pages that you printed to the science notebook.

Events in Science
Current Events
☻ Find a current events article related to the field of physics, and complete the article summary sheet found on p. 187 of the Appendix. Once you are done, add the sheet to the events section of your science notebook.

Hands-on Activity
Optional Hands-on
✖ Complete the power worksheet practice in the Appendix on p. 199. Add your work and answers to the science notebook.

Week 6 — Unit 3 (Hands-on Course) — 4-Day

Weekly Topic

→ This week will begin a look at work and machines.

	Day 1	Day 2	Day 3	Day 4
Textbook and Experiment	☐ Read *CK-12 Physical Science* Section 16.1.	☐ Read *CK-12 Physical Science* Section 16.2.	☐ Do the experiment "LEGO Balloon Car."	
Writing	☐ Add the vocabulary to the glossary section of your science notebook.	☐ Answer the assigned questions in the reading section of your science notebook.	☐ Record what you have done in the lab section of your science notebook.	☐ Complete the Events in Science assignment.

Supplies Needed

✓ **Experiment** – Several LEGO bricks, including wheels and axles, 3 Balloons

Other Notes

Week 6 — Unit 3 (Digital Course) — 3-Day

Weekly Topic

→ This week will begin a look at work and machines.

	Day 1	Day 2	Day 3
Textbook and Experiment	☐ Read *CK-12 Physical Science* Section 16.1.	☐ Read *CK-12 Physical Science* Section 16.2.	☐ Do the online lab "Energy of Combustion."
Writing	☐ Add the vocabulary to the glossary section of your science notebook.	☐ Answer the assigned questions in the reading section of your science notebook.	☐ Record what you have done in the lab section of your science notebook. ☐ Complete the Events in Science assignment.

Other Notes

Week 7 Notes - Work and Machines, Part 2

Textbook Assignments
Reading
- *CK-12 Physical Science* Sections 16.3, 16.4

Written
After you finish reading, answer questions #2–4, 6 in section 16.3 and questions #1, 3, 5 in section 16.4. File your work in the reading section of your science notebook. Then define the following terms in the glossary section of your science notebook:
- ☐ Compound machine
- ☐ Inclined plane
- ☐ Pulley
- ☐ Wedge

Unit Test
Take Unit Test #4 – Motion and Forces starting on *CK-12 Physical Science*, p. 346.

Experiment - Simple Machines
Materials
- ✓ Materials will vary based on what you choose to build.

Pre-Reading
A simple machine is a tool that you can use to help you to do work. In other words, simple machines make the task of lifting or moving an object easier. There are seven main types of simple machines.
- **Lever** – A rigid bar that is free to move around at a fixed point. Levers are frequently used to lift things (i.e., using a flat piece of metal to pry open a paint can, using a wheelbarrow to move dirt, or using a shovel to life a large rock).
- **Wheel and axle** – Two disks or cylinders, each with a different radius. The wheel and axle simple machine is often used to turn or rotate things (i.e., using a screwdriver to turn a screw or using the steering wheel of a vehicle).
- **Gears** – Toothed wheels that interlock in pairs; each one helps to drive the next. Gears are often used to consistently drive an object, such as a watch or clock.
- **Inclined plane** – A slanted surface that helps move objects up an incline. The inclined planes are typically ramps that can be used to move a heavy object from a lower level to a higher level.
- **Wedge** – A v-shaped object whose sides are two inclined planes. Wedges, like knives, axes, and zippers, make it easier to separate two objects.
- **Screw** – An inclined plane wrapped around a cylinder; this plane is also called the thread of the screw. These simple machines, including screws, nuts, and bolts, make it easier to drive an object into another object.
- **Pulley** – A rope that fits in the groove of a wheel. Pulleys make it easier to pull a heavy load directly upward.

In today's experiment, you are going to choose one of these types of simple machines to build.

Procedure

1. Choose one of the types of simple machines (lever, wheel and axle, gear, inclined plane, wedge, screw, or pulley) to build. See above for an explanation of each of the types of simple machines or visit the following website to get ideas:
   - https://elementalscience.com/blogs/science-activities/3-ideas-for-building-simple-machines-at-home
2. Construct the machine, and then test to see if it helps to perform work. For instance, if you choose to build a lever, test it by using a stack of books. First, pick up the stack on your own, then use the lever to pick up the stack. Which was easier to do?
3. Build as many as you have time to build.
4. Draw conclusions, and add them to the science notebook.

Science Notebook

- Write down a short description of what you did for the experiment and add the results. After that, add what you have learned from the experiment.

Online Lab - How Ramps Change Force and Work

Purpose

The purpose of this online lab is to compare the amount of work it takes to lift an object straight up in the air to pushing it up a ramp.

Pre-Reading

- Print and read the section of the workbook for the "How Ramps Change Force and Work" online lab.

Procedure

- Do the lab entitled "How Ramps Change Force and Work," and answer the questions as you work through the online lab.

Science Notebook

- Add the completed workbook pages that you printed to the science notebook.

Events in Science

Current Events

- Find a current events article related to the field of physics, and complete the article summary sheet found on p. 187 of the Appendix. Once you are done, add the sheet to the events section of your science notebook.

Hands-on Activity

Optional Hands-on

- Play the simple machines game from the Museum of Science and Industry in Chicago.
  - http://www.msichicago.org/play/simplemachines/

Week 7 — Unit 3 (Hands-on Course) — 4-Day

Weekly Topic

→ This week will wrap up a look at work and machines.

	Day 1	Day 2	Day 3	Day 4
Textbook and Experiment	☐ Read *CK-12 Physical Science* Section 16.3.	☐ Read *CK-12 Physical Science* Section 16.4.	☐ Do the experiment "Simple Machines."	☐ Do the optional hands-on science activity. ☐ Take Unit Test #4 – Motion and Forces starting on *CK-12 Physical Science*, p. 346.
Writing	☐ Add the vocabulary to the glossary section of your science notebook.	☐ Answer the assigned questions in the reading section of your science notebook.	☐ Record what you have done in the lab section of your science notebook.	☐ Complete the Events in Science assignment.

Supplies Needed

✓ **Experiment** – Materials will vary based on what you choose to build

Other Notes

Week 7 — Unit 3 (Digital Course) — 3-Day

Weekly Topic

→ This week will wrap up a look at work and machines.

	Day 1	Day 2	Day 3
Textbook and Experiment	☐ Read *CK-12 Physical Science* Section 16.3.	☐ Read *CK-12 Physical Science* Section 16.4. ☐ *(Optional)* Do the hands-on science activity.	☐ Do the online lab "How Ramps Change Force and Work." ☐ Take Unit Test #4 – Motion and Forces starting on *CK-12 Physical Science*, p. 346.
Writing	☐ Add the vocabulary to the glossary section of your science notebook.	☐ Answer the assigned questions in the reading section of your science notebook.	☐ Record what you have done in the lab section of your science notebook. ☐ Complete the Events in Science assignment.

Other Notes

Unit 3 - Answers

Week 2
Science Notebook Assignment (Experiment)
- 1. Resultant force = 0N, object is in balance
- 2. Resultant force = -2N, object will begin moving in the opposite direction
- 3. Resultant force = 8N, object will continue in the same direction
- 4. Resultant force = 0N, object is in balance

Week 4
Science Notebook Assignment (Experiment)
- 1. $a = 8 \text{ m/s}^2$
- 2. $F = 160{,}000 \text{ N}$
- 3. $m = 25 \text{ kg}$

Week 6
Science Notebook Assignment (Hands-on Activity)
- 1. 125 Watts
- 2. 77 Watts
- 3. 2236 Watts

Physical Science

Unit 4 – Energy and Electricity

Unit 4 – Overview of Study

Topics Covered

- Week 1: Introduction to Energy
- Week 2: Thermal Energy
- Week 3: Waves
- Week 4: Sound
- Week 5: Electromagnetic Radiation
- Week 6: Visible Light
- Week 7: Electricity, Part 1
- Week 8: Electricity, Part 2
- Week 9: Magnetism
- Week 10: Electromagnetism

Supplies Needed (for the Hands-on Course)

Week	Experiment	Optional Activity
1	Goldfish cracker, Small marshmallow, Piece of lettuce, Piece of bacon fat, Aluminum pan, Matches, Safety glasses, Bucket of water	PVC pipe, Propeller blade, DC motor, Electrical tape, Wire
2	Foil, Black construction paper, Small cardboard box (about 9" x 12"), Plastic wrap, Tape, Marshmallow, Small glass dish (one that will fit inside the box)	3 Glasses, Red and blue food coloring, Ice cold water, Hot water
3	Shallow glass bowl or cup, Water, Music Player, Measuring tape	No supplies needed
4	Plastic jar or small flowerpot, Piece of latex material large enough to cover the lid of your jar (like the kind used for exercise bands), 1" Plastic tubing, Rubber band, Air-dry clay, Salt	No supplies needed
5	4 Pencils, 4 Clear glasses, Water, Oil, Alcohol, Corn syrup	Mirror, Flashlight, Dark room
6	Thin cardboard, Red, blue, and yellow paint, 6 Rubber bands, Hole punch	Materials will vary
7	Styrofoam tray or plate, Aluminum pan or pie-plate, Wool, Plastic tongs, Pin	Plastic comb, wool, water
8	Light bulb, Copper wire, D-battery, Electrical tape, 2 Alligator clips, Organic material, such as a pickle, lemon slice, cheese, bread, or a leaf	No supplies needed
9	2 Different types of magnets, such as a horseshoe magnet and a neodymium magnet, Paper clips (20 to 30), Paper, Cardboard, Thick books	2 Bar magnets, Iron filings
10	D-battery, Insulated copper wire – about 3 ft (1 m), 2 to 3 in (5 to 8 cm) Nail, Electrical tape, Iron filings, Paper	Battery, 2 Safety pins, Rubber band, Bobbin, Copper wire

Week 1 Notes - Introduction to Energy

Textbook Assignments
Reading
📖 *CK-12 Physical Science* Sections 17.1, 17.2, 17.3

Written
After you finish reading, answer questions #1–4 in section 17.1, questions #1–4 in section 17.2, and questions #2, 4, 6, 8 in section 17.3. File your work in the reading section of your science notebook. Then define the following terms in the glossary section of your science notebook:

- ☐ Chemical energy
- ☐ Conservation
- ☐ Electromagnetic energy
- ☐ Energy conversion
- ☐ Fossil fuel
- ☐ Mechanical energy
- ☐ Natural resource
- ☐ Nonrenewable resource
- ☐ Potential energy
- ☐ Renewable resource
- ☐ Thermal energy

Experiment - Energy in Food
Materials
- ✓ Goldfish cracker
- ✓ Small marshmallow
- ✓ Piece of lettuce
- ✓ Piece of bacon fat
- ✓ Aluminum pan
- ✓ Matches
- ✓ Safety glasses
- ✓ Bucket of water

Pre-Reading
As we break down the food we eat, it releases energy in the form of calories. Some of the chemicals in food are broken down quickly, releasing a burst of energy. Other chemicals take a bit more work to break down so they release energy slowly over time. These reactions work together to fuel our bodies' processes over time. In today's experiment, you will be testing different types of food to see how much energy they release.

Procedure
1. Head outside and set the aluminum pan on a concrete or tile surface or inside a grill.
   - **Caution:** Be sure to do this experiment in a well-ventilated area with a fire extinguisher close at hand.
2. Place the cracker in the aluminum pan and have an adult use the match to light the cracker. Time how long it takes for the cracker to burn and write this on your science notebook.
3. Use the bucket of water to extinguish any smoldering embers. Then, clean and dry the aluminum pan.
4. Now, repeat the procedure from step 2 and 3 for the marshmallow, lettuce, and bacon fat. (**Note—***Make sure that your samples are similar in size, so that your results are more accurate.*)

5. Draw conclusions and add them to the science notebook.

Explanation

You should see that the marshmallow burned the quickest, followed by the lettuce, the goldfish cracker, and finally the bacon. The marshmallow has a quick sugar-rush of energy, which burns and then quickly dies out. The lettuce has fiber and a bit of sugar, which burns slower than the marshmallow. However, since lettuce is mostly water, it doesn't burn all that long. The goldfish cracker has a mixture of sugar, starch, and fiber, which keeps it burning longer than the two previous samples. The bacon fat is full of lipids and proteins that burn for a long time. Just like the food we eat, vegetables give us the fuel our bodies needs, but we have to eat a lot of bulk to survive on only vegetables. Sweets, like marshmallows, give our bodies a quick punch of energy that doesn't last very long. Carbohydrates rich in fiber fuel our bodies for a much longer period of time than sugar. Fats and proteins provide our bodies with a long-lasting source of energy. Since our bodies need more than just fuel, it is best to have a balance of fruits and vegetables, carbohydrates, along with fats and proteins.

Science Notebook

☞ Write down a short description of what you did for the experiment and add a chart with the results. After that, add what you have learned from the experiment.

Online Lab - Potential Energy to Kinetic Energy

Purpose

The purpose of this online lab is to learn how to use graphs to describe the motion of objects.

Pre-Reading

✍ Print and read the section of the workbook for the "Potential Energy to Kinetic Energy" online lab.

Procedure

✓ Do the lab entitled "Potential Energy to Kinetic Energy," and answer the questions as you work through the online lab.

Science Notebook

☞ Add the completed workbook pages that you printed to the science notebook.

Events in Science

Current Events

🕓 Find a current events article related to the field of physics, and complete the article summary sheet found on p. 187 of the Appendix. Once you are done, add the sheet to the events section of your science notebook.

Hands-on Activity
Optional Hands-on

✂ Make your own wind turbine from home! You will need PVC pipe, a propeller blade, a DC motor, electrical tape, and wire. This is a bit of an ambitious project, but you can see how to build your own here:
💻 https://www.youtube.com/watch?v=YY1oCNhD8_0

Week 1	Unit 4 (Hands-on Course)			4-Day
Weekly Topic				
→ This week will be an introduction to energy.				
	Day 1	**Day 2**	**Day 3**	**Day 4**
Textbook and Experiment	☐ Read *CK-12 Physical Science* Section 17.1 and 17.2.	☐ Read *CK-12 Physical Science* Section 17.3.	☐ Do the experiment "Energy in Food."	☐ Do the optional hands-on science activity.
Writing	☐ Add the vocabulary to the glossary section of your science notebook.	☐ Answer the assigned questions in the reading section of your science notebook.	☐ Record what you have done in the lab section of your science notebook.	☐ Complete the Events in Science assignment.
Supplies Needed				
✓ Experiment – Goldfish cracker, Small marshmallow, Piece of lettuce, Piece of bacon fat, Aluminum pan, Matches Safety glasses, Bucket of water
✓ *(Optional)* PVC pipe, Propeller blade, DC motor, Electrical tape, Wire				
Other Notes				

Week 1	Unit 4 (Digital Course)	3-Day

Weekly Topic

→ This week will be an introduction to energy.

	Day 1	Day 2	Day 3
Textbook and Experiment	☐ Read *CK-12 Physical Science* Section 17.1 and 17.2.	☐ Read *CK-12 Physical Science* Section 17.3.	☐ Do the online lab "Potential Energy to Kinetic Energy." ☐ *(Optional)* Do the hands-on science activity.
Writing	☐ Add the vocabulary to the glossary section of your science notebook.	☐ Answer the assigned questions in the reading section of your science notebook.	☐ Record what you have done in the lab section of your science notebook. ☐ Complete the Events in Science assignment.

Other Notes

Week 2 Notes - Thermal Energy

Textbook Assignments
Reading
- *CK-12 Physical Science* Sections 18.1, 18.2, 18.3

Written
After you finish reading, answer questions #1–4 in section 18.1, questions #1–5 in section 18.2, and questions #1–4 in section 18.3. File your work in the reading section of your science notebook. Then define the following terms in the glossary section of your science notebook:

- [ ] Combustion engine
- [ ] Conduction
- [ ] Convection
- [ ] Convection Current
- [ ] Heat
- [ ] Refrigerant
- [ ] Specific Heat
- [ ] Thermal Conductor
- [ ] Thermal Insulator

Experiment - Solar Oven
Materials
- ✓ Foil
- ✓ Black construction paper
- ✓ Small cardboard box (about 9" by 12")
- ✓ Plastic wrap
- ✓ Tape
- ✓ Marshmallow
- ✓ Small glass dish (one that will fit inside the box)

Pre-Reading
On Earth, we have many different sources of energy. Some of these sources are not renewable, such as fossil fuels like oil and coal. Some of the sources are renewable, like wind, waves, and the sun. In this experiment, you are going to build a device that will allow you to harness the power of the sun to cook a marshmallow.

Procedure
1. Place the black construction paper on the bottom of your box. This will serve to absorb the sunlight.
2. Tape the flaps down. Then, cover the sides with the foil so that the shiny side is facing out. The foil will help to reflect the sunlight and focus it on cooking the marshmallow.
3. Put the glass dish in the center of the bottom of the box and set the marshmallow on it. Cover the top box with plastic wrap; this will help to trap the heat.
4. Now, head outside and set your solar oven in direct sunlight. Observe what happens.
5. Draw conclusions and add them to the science notebook.

Troubleshooting
Be sure that the solar oven is in direct sunlight at the peak of the day (between 11am and 3pm).

Explanation
The purpose of this experiment was to give you a chance to see how thermal energy from the sun can be used as an energy source. If you were able to cook your marshmallow, you have accomplished this purpose.

Science Notebook
☞ Write down a short description of what you did for the experiment and add the results. After that, add what you have learned from the experiment.

Online Lab - Comparing Thermal Energy in Metals
Purpose
The purpose of this online lab is to investigage the relationship between different types of materials and the amount of energy transferred to water as meansure by a temperature change.

Pre-Reading
✍ Print and read the section of the workbook for the "Comparing Thermal Energy in Metals" online lab.

Procedure
✓ Do the lab entitled "Comparing Thermal Energy in Metals," and answer the questions as you work through the online lab.

Science Notebook
☞ Add the completed workbook pages that you printed to the science notebook.

Events in Science
Current Events
🕐 Find a current events article related to the field of physics, and complete the article summary sheet found on p. 187 of the Appendix. Once you are done, add the sheet to the events section of your science notebook.

Hands-on Activity
Optional Hands-on
✂ Explore convection of heat. You will need three glasses, red and blue food coloring, ice cold water, and hot water. (**Note**—*Do not to handle the hot water without the proper protection.*) Begin by filling one of the glasses two thirds of the way with cold water and add three drops of blue food coloring. Observe what happens and record the time it takes for the food coloring to completely mix. Next, fill the other glass two thirds of the way with hot water and add two drops of food coloring. Observe what happens and record the time it takes for the food coloring to completely mix. Then pour one quarter of the hot water very slowly into the cold water, observe what happens, and record it. After 30 minutes, check the glass once more to observe and record what has happened.

Physical Science Unit 4 - Week 2

- **Explanation:** You should see that it took about three times as long for the food coloring to fully mix into the cold water than in the hot water. You should also see that the hot red water layer "floats" on top of the cold blue water layer. After 30 minutes, you should see that the water is now purple because the two layers have mixed. This is because hot water molecules move faster than cold water molecules which causes the food coloring in the hot water to mix in much quicker. Also, hot water molecules are more spread out, making the same volume of hot water to be less dense than cold water. This allows the hot water to appear to float on the surface of the cold water. As the hot water cools and the cool water warms up, the two mix turning the water purple.

Week 2	Unit 4 (Hands-on Course)			4-Day
Weekly Topic				
→ This week will be a look at thermal energy.				
	Day 1	**Day 2**	**Day 3**	**Day 4**
Textbook and Experiment	❏ Read *CK-12 Physical Science* Section 18.1 and 18.2.	❏ Read *CK-12 Physical Science* Section 18.3.	❏ Do the experiment "Solar Oven."	❏ Do the optional hands-on science activity.
Writing	❏ Add the vocabulary to the glossary section of your science notebook.	❏ Answer the assigned questions in the reading section of your science notebook.	❏ Record what you have done in the lab section of your science notebook.	❏ Complete the Events in Science assignment.

Supplies Needed

✓ **Experiment** – Foil, Black construction paper, Small cardboard box (about 9" by 12"), Plastic wrap, Tape, Marshmallow, Small glass dish (one that will fit inside the box)
✓ *(Optional)* 3 Glasses, Red and blue food coloring, Ice cold water, Hot water

Other Notes

Week 2 — Unit 4 (Digital Course) — 3-Day

Weekly Topic

→ This week will be a look at thermal energy.

	Day 1	Day 2	Day 3
Textbook and Experiment	❏ Read *CK-12 Physical Science* Section 18.1 and 18.2.	❏ Read *CK-12 Physical Science* Section 18.3.	❏ Do the online lab "Comparing Thermal Energy in Metals." ❏ *(Optional)* Do the hands-on science activity.
Writing	❏ Add the vocabulary to the glossary section of your science notebook.	❏ Answer the assigned questions in the reading section of your science notebook.	❏ Record what you have done in the lab section of your science notebook. ❏ Complete the Events in Science assignment.

Other Notes

Week 3 Notes - Waves

Textbook Assignments
Reading
- 📖 *CK-12 Physical Science* Section 19.1, 19.2, 19.3

Written
After you finish reading, answer questions #1–6 in section 19.1, questions #1–4 in section 19.2, and questions #1–5 in section 19.3. File your work in the reading section of your science notebook. Then define the following terms in the glossary section of your science notebook:

- ☐ Diffraction
- ☐ Hertz
- ☐ Longitudinal wave
- ☐ Mechanical wave
- ☐ Reflection
- ☐ Refraction
- ☐ Standing wave
- ☐ Surface wave
- ☐ Transverse wave
- ☐ Wave amplitude
- ☐ Wave frequency
- ☐ Wave interference
- ☐ Wave speed
- ☐ Wavelength

Experiment - Distance and Sound
Materials
- ✓ Shallow glass bowl or cup
- ✓ Water
- ✓ Music Player
- ✓ Measuring tape

Pre-Reading
As a mechanical wave is produced when a source of energy creates a vibration that travels through a medium such as air or water. A vibration is basically repeated back-and-forth motion, and it can be seen on the surface of a medium such as water. As the sound waves move through the water, a vibration visually disturbs the surface. The greater the strength of the sound wave the more the water is disturbed. In today's experiment, you are going test if the distance from the energy source makes a different in the rate of vibrations that are transferred through the water.

Procedure
1. Fill the bowl with water and set it right next to the speaker of the music player. Choose a fast selection of music and play it at a relatively loud volume. Observe the changes in the water in the bowl and sketch what you see.
2. Now, move the player 12 in (30 cm) away from the bowl. Play the same selection of music once more. Observe the changes in the water in the bowl and sketch what you see.
3. Then move the player 24 in (60 cm) away from the bowl. Play the same selection of music again. Observe the changes in the water in the bowl, and sketch what you see.
4. Finally, move the player 36 in (90 cm) away from the bowl. Play the same selection of music for the final time. Observe the changes in the water in the bowl and sketch what you see.

5. Draw conclusions and add them to the science notebook.

Explanation
You should see that the farther the music player is from the bowl the less the water is disturbed by the vibrations. As sound travels through the air, it exerts pressure on the surrounding medium. Over time, the pressure the waves exert is dampened, and the amplitude, or loudness, of the sound wave is decreased. Eventually, the sound waves are not felt, or heard, anymore. In the experiment, you saw this effect in action. As the music player was moved further away from the water bowl, the amplitude of the sound waves decreased, which caused the overall effect of the vibrations in the water to be decreased.

Science Notebook
- Write down a short description of what you did for the experiment and add the results. After that, add what you have learned from the experiment.
- Complete the mechanical wave worksheet practice in the Appendix on p. 200. Add your work and answers to the science notebook.

Online Lab - Transmission and Reflection of Light
Purpose
The purpose of this online lab is to study the properties of light when intercepting transparent and opaque objects and to examine reflections of images off plane mirrors from different angles.

Pre-Reading
- Print and read the section of the workbook for the "Transmission and Reflection of Light" online lab.

Procedure
- Do the lab entitled "Transmission and Reflection of Light," and answer the questions as you work through the online lab.

Science Notebook
- Add the completed workbook pages that you printed to the science notebook.

Events in Science
Current Events
- Find a current events article related to the field of physics, and complete the article summary sheet found on p. 187 of the Appendix. Once you are done, add the sheet to the events section of your science notebook.

Hands-on Activity
Optional Hands-on
- Do the following online sound simulation from PhET:
  http://phet.colorado.edu/en/simulation/sound

Week 3	Unit 4 (Hands-on Course)			4-Day
Weekly Topic				
→ This week will look at waves.				
	Day 1	**Day 2**	**Day 3**	**Day 4**
Textbook and Experiment	☐ Read *CK-12 Physical Science* Section 19.1 and 19.2.	☐ Read *CK-12 Physical Science* Section 19.3.	☐ Do the experiment "Distance and Sound."	☐ Do the optional hands-on science activity.
Writing	☐ Add the vocabulary to the glossary section of your science notebook.	☐ Answer the assigned questions in the reading section of your science notebook.	☐ Record what you have done in the lab section of your science notebook.	☐ Complete the Events in Science assignment.

Supplies Needed

✓ **Experiment** – Shallow glass bowl or cup, Water, Music Player, Measuring tape

Other Notes

Week 3	Unit 4 (Digital Course)		3-Day
Weekly Topic			
→ This week will look at waves.			

	Day 1	Day 2	Day 3
Textbook and Experiment	☐ Read *CK-12 Physical Science* Section 19.1 and 19.2.	☐ Read *CK-12 Physical Science* Section 19.3.	☐ Do the online lab "Transmission and Reflection of Light." ☐ *(Optional)* Do the hands-on science activity.
Writing	☐ Add the vocabulary to the glossary section of your science notebook.	☐ Answer the assigned questions in the reading section of your science notebook.	☐ Record what you have done in the lab section of your science notebook. ☐ Complete the Events in Science assignment.

Other Notes

Week 4 Notes - Sound

Textbook Assignments
Reading
📖 *CK-12 Physical Science* Sections 20.1, 20.2, 20.3

Written
After you finish reading, answer questions #2, 4, 6 in section 20.1, questions #1–5 in section 20.2, and questions #1–6 in section 20.3. File your work in the reading section of your science notebook. Then define the following terms in the glossary section of your science notebook:

- ☐ Decibel
- ☐ Doppler effect
- ☐ Infrasound
- ☐ Intensity
- ☐ Resonance
- ☐ Sonar
- ☐ Ultrasound

Experiment - Tonoscope
Materials
- ✓ Plastic jar or small flowerpot
- ✓ A piece of latex material large enough to cover the lid of your jar (like the kind used for exercise bands)
- ✓ 1" plastic tubing
- ✓ Rubber band
- ✓ Air-dry clay
- ✓ Salt

Pre-Reading
A tonoscope is an acoustically device that allows you to see patterns in salt created by the sound of your voice. It was originally designed by Hans Jeny to explore wave behavior and patterns. In today's experiment, you are going to make your own tonoscope.

Procedure
1. Cut a hole in the lower half of the plastic jar large enough to fit the tubing. Insert the tubing into the hole you just made. Then pack a bit of clay around the edges so that air won't leak out.
2. Next, stretch the latex material over the opening of the jar and use the rubber band to secure it in place. Sprinkle a bit of salt over the rubber material.
3. Hum, sing, or speak through the tubing and watch what happens to the salt. Change the pitch of your voice and observe any changes. Write your observations on your science notebook.
4. Draw conclusions and add them to the science notebook.

Explanation
You should see the salt crystals vibrate and move when they hum. As they change the pitch of their hum, the salt crystals should move to make different patterns such as rings or waves. Your voice creates vibrations when you hum or talk into the tonoscope. As you change the pitch of your voice, you are able to tune into the different natural frequencies

of the membrane causing it to resonate. Because of the variations in the latex, some parts of the membrane resonate while other don't. The parts that remain still are known as nodal points and the salt tends to move toward and collect at these points creating patterns on the tonoscope lid.

Science Notebook
☞ Write down a short description of what you did for the experiment and add the results. Finally, add what you have learned from the experiment.

Online Lab
☞ There is no online lab scheduled for this week.

Events in Science
Current Events
🕓 Find a current events article related to the field of physics, and complete the article summary sheet found on p. 187 of the Appendix. Once you are done, add the sheet to the events section of your science notebook.

Hands-on Activity
Optional Hands-on
✂ Watch the following video about the Doppler Effect:
🖥 https://www.youtube.com/watch?v=h4OnBYrbCjY

Week 4	Unit 4 (Hands-on Course)			4-Day
Weekly Topic				
→ This week will look at sound.				
	Day 1	**Day 2**	**Day 3**	**Day 4**
Textbook and Experiment	☐ Read *CK-12 Physical Science* Section 20.1 and 20.2.	☐ Read *CK-12 Physical Science* Section 20.3.	☐ Do the experiment "Tonoscope."	☐ Do the optional hands-on science activity.
Writing	☐ Add the vocabulary to the glossary section of your science notebook.	☐ Answer the assigned questions in the reading section of your science notebook.	☐ Record what you have done in the lab section of your science notebook.	☐ Complete the Events in Science assignment.
Supplies Needed				
✓ **Experiment** – Plastic jar or small flowerpot, Piece of latex material large enough to cover the lid of your jar (like the kind used for exercise bands), 1" Plastic tubing, Rubber band, Air-dry clay, Salt				
Other Notes				

Week 4	Unit 4 (Digital Course)		3-Day
Weekly Topic			
→ This week will look at sound.			
	Day 1	**Day 2**	**Day 3**
Textbook and Experiment	☐ Read *CK-12 Physical Science* Section 20.1 and 20.2.	☐ Read *CK-12 Physical Science* Section 20.3.	☐ *(Optional)* Do the hands-on science activity.
Writing	☐ Add the vocabulary to the glossary section of your science notebook.	☐ Answer the assigned questions in the reading section of your science notebook.	☐ Complete the Events in Science assignment.
Other Notes			

Week 5 Notes - Electromagnetic Radiation

Textbook Assignments
Reading
📖 *CK-12 Physical Science* Section 21.1, 21.2, 21.3

Written
After you finish reading, answer questions #1–5 in section 21.1, questions #1–4 in section 21.2, and questions #1, 3, 5, 7 in section 21.3. File your work in the reading section of your science notebook. Then define the following terms in the glossary section of your science notebook:

- ☐ Electromagnetic radiation
- ☐ Electromagnetic spectrum
- ☐ Electromagnetic wave
- ☐ Gamma ray
- ☐ Microwave
- ☐ Photon
- ☐ Radio wave
- ☐ Speed of Light
- ☐ Ultraviolet light
- ☐ Visible light
- ☐ X-ray

Experiment - Reflection and Refraction
Materials
- ✓ 4 Pencils
- ✓ 4 Clear glasses
- ✓ Water
- ✓ Oil
- ✓ Alcohol
- ✓ Corn syrup

Pre-Reading
🕰 Light typically travels in a straight line. However, as it passes through different media, it bends slightly. This phenomenon is known as refraction, and it occurs because light travels at different speeds through different media. In today's experiment, you are going to test how different liquids refract the light from a pencil in different liquids.

Procedure
1. Label the cups #1 through #4 and place a pencil in each cup. Observe how the pencil looks in the cup without any liquid.
2. Now add ½ cup (about 120 mL) of water to cup #1, add ½ cup (about 120 mL) of oil to cup #2, ½ cup (about 120 mL) of alcohol to cup #3, and ½ cup (about 120 mL) of corn syrup to cup #4.
3. Observe what happens to the pencils in the glasses and write your observations down on your science notebook.
4. Draw conclusions and add them to the science notebook.

Explanation
The pencil should not appear to be split in the cup by itself. The pencils should appear to be split by all four of the liquids. The degree of split will be greater in the oil and corn syrup. The refractive index of a material is a number that measures how much a light beam bends in

a given material. The refractive index of water is 1.3, alcohol is 1.35, oil is 1.5, and corn syrup is 1.4. Each of the materials bends the reflected light of the pencil, causing the image to distort and look like the pencil is split. The oil and the corn syrup appeared to split the pencil farther because their reflective indexes are higher than those of water and alcohol.

Science Notebook
- Write down a short description of what you did for the experiment and add the results. After that, add what you have learned from the experiment.
- Complete the electromagnetic wave worksheet practice in the Appendix on p. 201. Add your work and answers to the science notebook.

Online Lab - Reflection and Refraction of Light
Purpose
The purpose of this online lab is to compare reflections off flat and curved mirrors and study the refraction of light through a lens.

Pre-Reading
- Print and read the section of the workbook for the "Reflection and Refraction of Light" online lab.

Procedure
- Do the lab entitled "Reflection and Refraction of Light," and answer the questions as you work through the online lab.

Science Notebook
- Add the completed workbook pages that you printed to the science notebook.

Events in Science
Current Events
- Find a current events article related to the field of physics, and complete the article summary sheet found on p. 187 of the Appendix. Once you are done, add the sheet to the events section of your science notebook.

Hands-on Activity
Optional Hands-on
- Observe reflection using a mirror, flashlight, and dark room. Go into the dark room, shine the flashlight at the mirror, and observe where the beam of light ends up. (You should see that the beam of light is reflected onto the opposite wall.)
- Watch the following video from Khan Academy on Young's double-slit experiment, which shows how light behaves like a wave.
  - https://www.khanacademy.org/science/physics/light-waves/interference-of-light-waves/v/youngs-double-split-part-1

Week 5	Unit 4 (Hands-on Course)			4-Day
Weekly Topic				
→ This week will look at electromagnetic radation.				
	Day 1	**Day 2**	**Day 3**	**Day 4**
Textbook and Experiment	☐ Read *CK-12 Physical Science* Section 21.1 and 21.2.	☐ Read *CK-12 Physical Science* Section 21.3.	☐ Do the experiment "Reflection and Refraction."	☐ Do the optional hands-on science activity.
Writing	☐ Add the vocabulary to the glossary section of your science notebook.	☐ Answer the assigned questions in the reading section of your science notebook.	☐ Record what you have done in the lab section of your science notebook.	☐ Complete the Events in Science assignment.

Supplies Needed

- ✓ **Experiment** – 4 Pencils, 4 Clear glasses, Water, Oil, Alcohol, Corn syrup
- ✓ *(Optional)* Mirror, Flashlight, Dark room

Other Notes

Week 5 — Unit 4 (Digital Course) — 3-Day

Weekly Topic

→ This week will look at electromagnetic radation.

	Day 1	Day 2	Day 3
Textbook and Experiment	☐ Read *CK-12 Physical Science* Section 21.1 and 21.2.	☐ Read *CK-12 Physical Science* Section 21.3.	☐ Do the online lab "Reflection and Refraction of Light." ☐ *(Optional)* Do the hands-on science activity.
Writing	☐ Add the vocabulary to the glossary section of your science notebook.	☐ Answer the assigned questions in the reading section of your science notebook.	☐ Record what you have done in the lab section of your science notebook. ☐ Complete the Events in Science assignment.

Other Notes

Week 6 Notes - Visible Light

Textbook Assignments
Reading
📖 *CK-12 Physical Science* Sections 22.1, 22.2, 22.3
Written
After you finish reading, answer questions #1–5 in section 22.1, questions #1, 3, 5, 7 in section 22.2, and questions #1–6 in section 22.3. File your work in the reading section of your science notebook. Then define the following terms in the glossary section of your science notebook:

- ☐ Absorption
- ☐ Concave
- ☐ Convex
- ☐ Incandescence
- ☐ Law of reflection
- ☐ Opaque
- ☐ Optics
- ☐ Scattering
- ☐ Translucent
- ☐ Transmission
- ☐ Transparent

Experiment - Color Change
Materials
- ✓ Thin cardboard
- ✓ Red, blue, and yellow paint
- ✓ 6 Rubber bands
- ✓ Hole punch

Pre-Reading
✎ Visible light is made up of a spectrum of wavelengths. When we see color, we are actually see the wavelength(s) of light that are reflected by the object. Light receptors in our eyes receive the reflected wavelengths, and then our brain translated it into the sensation of color. In today's experiment, you are going to try to trick your brain into seeing different colors than are actually there.

Procedure
1. Cut out six circles of the same size from the cardboard. Paint two of them red, two of them yellow, and two of them blue. Set them aside to dry.
2. Once they are dry, glue one of the red disks to one of the yellow disks, one of the red disks to one of the blue disks, and one of the blue disks to one of the yellow disks. Then punch holes on opposite sides of the three disks and tie a rubber band into each hole.
3. Take the red/blue disk, twist the rubber bands, and then stretch them out so that the disk spins quickly. Observe what color you see and write it on your science notebook.
4. Repeat step 4 with the red/yellow and blue/yellow disks.
5. Draw conclusions and add them to the science notebook.

Troubleshooting
If you can still see both colors, the disks are not spinning fast enough.

Explanation

When you spin the red/blue disk, you will see purple. When you spin the red/yellow disk, you will see orange. When you spin the blue/yellow disk, you will see green. We know that when we mix the primary colors of red and blue we make purple. Obviously, the colors on the two opposite disks are not physically mixing. Instead, the disks are moving so fast that your brain is unable to process the individual colors of the disks. The brain creates a shortcut and combines the two wavelengths together to create the perception of seeing the secondary color.

Science Notebook

☞ Write down a short description of what you did for the experiment and add the results. After that, add what you have learned from the experiment.

Online Lab - Color in Light

Purpose

The purpose of this online lab is to observe light color addition and subtraction through filters, receptors in the eye, different types of color blindness, and prisms.

Pre-Reading

✐ Print and read the section of the workbook for the "Color in Light" online lab.

Procedure

✓ Do the lab entitled "Color in Light," and answer the questions as you work through the online lab.

Science Notebook

☞ Add the completed workbook pages that you printed to the science notebook.

Events in Science

Current Events

⏱ Find a current events article related to the field of physics, and complete the article summary sheet found on p. 187 of the Appendix. Once you are done, add the sheet to the events section of your science notebook.

Hands-on Activity

Optional Hands-on

✂ Make your own telescope using the directions from the following website:
💻 http://www.space.com/24114-how-to-build-a-telescope-science-fair-projects.html

Physical Science Unit 4 - Week 6

Week 6 — Unit 4 (Hands-on Course) — 4-Day

Weekly Topic

→ This week will look at visible light.

	Day 1	Day 2	Day 3	Day 4
Textbook and Experiment	☐ Read *CK-12 Physical Science* Section 22.1 and 22.2.	☐ Read *CK-12 Physical Science* Section 22.3.	☐ Do the experiment "Color Change."	☐ Do the optional hands-on science activity.
Writing	☐ Add the vocabulary to the glossary section of your science notebook.	☐ Answer the assigned questions in the reading section of your science notebook.	☐ Record what you have done in the lab section of your science notebook.	☐ Complete the Events in Science assignment.

Supplies Needed

✓ **Experiment** – Thin cardboard, Red, blue, and yellow paint, 6 Rubber bands, Hole punch
✓ *(Optional)* Materials will vary

Other Notes

Week 6	Unit 4 (Digital Course)		3-Day
Weekly Topic			
→ This week will look at visible light.			
	Day 1	**Day 2**	**Day 3**
Textbook and Experiment	☐ Read *CK-12 Physical Science* Section 22.1 and 22.2.	☐ Read *CK-12 Physical Science* Section 22.3.	☐ Do the online lab "Colors in Light." ☐ *(Optional)* Do the hands-on science activity.
Writing	☐ Add the vocabulary to the glossary section of your science notebook.	☐ Answer the assigned questions in the reading section of your science notebook.	☐ Record what you have done in the lab section of your science notebook. ☐ Complete the Events in Science assignment.
Other Notes			

Week 7 Notes - Electricity, Part 1

Textbook Assignments
Reading
- *CK-12 Physical Science* Sections 23.1, 23.2

Written
After you finish reading, answer questions #2–4, 6 in section 23.1 and questions #2, 4, 6, 8 in section 23.2. File your work in the reading section of your science notebook. Then define the following terms in the glossary section of your science notebook:

- [ ] Alternating current
- [ ] Direct current
- [ ] Electric charge
- [ ] Electric current
- [ ] Electric field
- [ ] Electric force
- [ ] Law of conservation of energy
- [ ] Ohm's law
- [ ] Static discharge
- [ ] Static electricity

Experiment - Electrical Charge Transfer
Materials
- ✓ Styrofoam tray or plate
- ✓ Aluminum pan or pie-plate
- ✓ Wool
- ✓ Plastic tongs
- ✓ Pin

Pre-Reading
An electrical charge is produced when a subatomic particle within an object has a shortage or excess of electrons. A shortage of electrons produces a positive electrical charge, and an excess of electrons produces a negative electrical charge. The two types of electrical charges are attracted to one another, and electricity takes advantage of this principle. In today's experiment, you are going to test whether you can transfer an electrical charge from one object to another.

Procedure
1. Rub the Styrofoam tray (or plate) vigorously with a piece of wool for about two minutes. Set the tray face down on a smooth surface.
2. Use the plastic tongs to set the aluminum pan (or pie-plate) on top of the Styrofoam tray (or plate). Then gently slide it back and forth several times.
3. Now, use the tongs to pick up the pin. With your hands on the plastic tongs, move the pin close to the pan and observe what happens.
4. Draw conclusions and add them to the science notebook.

Troubleshooting
If you do not see a spark, you may need to rub the Styrofoam more vigorously with the wool. You can also try rubbing the tray against your hair.

Explanation
You should hear and see a spark as the pin gets within a half-inch or less from the pan.

By rubbing the Styrofoam tray with the wool, you produced a static electrical charge through friction. This static electric charge is then transferred to the aluminum pan through contact. When the pin came in close contact with the aluminum pan, the static charge is transferred by induction from the pan to the pin. This process gave you a firsthand look at the three methods of transferring electrical charge: friction, contact, and induction.

Science Notebook

☞ Write down a short description of what you did for the experiment and add the results. After that, add what you have learned from the experiment.

Online Lab

☞ There is no online lab scheduled for this week.

Events in Science
Current Events

🕒 Find a current events article related to the field of physics, and complete the article summary sheet found on p. 187 of the Appendix. Once you are done, add the sheet to the events section of your science notebook.

Hands-on Activity
Optional Hands-on

✂ Test the effect of electrical charge on a stream of water. Begin by rubbing a plastic comb with wool. Then place the comb near a stream of water and observe what happens.
- **Explanation:** You should see that the stream of water is attracted to the comb. This is because the comb has been charged by the wool, and it now can attract the slightly polar stream of water.

Week 7	Unit 4 (Hands-on Course)			4-Day
Weekly Topic				

→ This week will finish a look at electricity.

	Day 1	Day 2	Day 3	Day 4
Textbook and Experiment	☐ Read *CK-12 Physical Science* Section 23.1.	☐ Read *CK-12 Physical Science* Section 23.2.	☐ Do the experiment "Electrical Charge Transfer."	☐ Do the optional hands-on science activity.
Writing	☐ Add the vocabulary to the glossary section of your science notebook.	☐ Answer the assigned questions in the reading section of your science notebook.	☐ Record what you have done in the lab section of your science notebook.	☐ Complete the Events in Science assignment.

Supplies Needed

✓ **Experiment** – Styrofoam tray or plate, Aluminum pan or pie-plate, Wool, Plastic tongs, Pin
✓ *(Optional)* Plastic comb, wool, water

Other Notes

Week 7	Unit 4 (Digital Course)		3-Day

Weekly Topic

→ This week will finish a look at electricity.

	Day 1	Day 2	Day 3
Textbook and Experiment	☐ Read *CK-12 Physical Science* Section 23.1.	☐ Read *CK-12 Physical Science* Section 23.2.	☐ *(Optional)* Do the hands-on science activity.
Writing	☐ Add the vocabulary to the glossary section of your science notebook.	☐ Answer the assigned questions in the reading section of your science notebook.	☐ Complete the Events in Science assignment.

Other Notes

Physical Science Unit 4 - Week 7

Week 8 Notes - Electricity, Part 2

Textbook Assignments
Reading
📖 *CK-12 Physical Science* Sections 23.3, 23.4

Written
After you finish reading, answer questions #1, 3, 5, 7 in section 23.3 and questions #1–6 in section 23.4. File your work in the reading section of your science notebook. Then define the following terms in the glossary section of your science notebook:

- ☐ Electric circuit
- ☐ Electric power
- ☐ Electronics
- ☐ Parallel circuit
- ☐ Semiconductor
- ☐ Series circuit

Experiment - Conduction
Materials
- ✓ Light bulb
- ✓ Copper wire
- ✓ D-battery
- ✓ Electrical tape
- ✓ 2 Alligator clips
- ✓ Organic material, such as a pickle, lemon slice, cheese, bread, or a leaf

Pre-Reading
✍ A conductor is a material through which electrical charge can easily flow while an insulator is a material through which electrical charge cannot easily flow. When we want to move electricity from one place to another, we use copper wire, which easily conducts electrical charge. But what if we could use other renewable organic materials, like pickles, to conduct electricity? In today's lab, you are going to test whether different organic compounds can conduct electricity.

Procedure
1. Begin by cutting the wire into three lengths. Then take one of the wires and attach an alligator clip to one end. Take the next wire and attach one end to the other alligator clip. Then wrap the other end once around the base of a light bulb.
2. Now, take the two wires with bare ends and attach those to the two terminals of the battery using the electrical tape. (See the diagram above to visually check your electrical circuit.)
3. The electrical circuit is now ready for testing. Simply clip the alligator clips on either side of your sample, hold the top of the bulb, and touch the base of the light bulb to the end of the wire coming from the battery to complete the circuit. If the light bulb lights up, the material conducts electricity; if it does not, the sample does not conduct electricity.

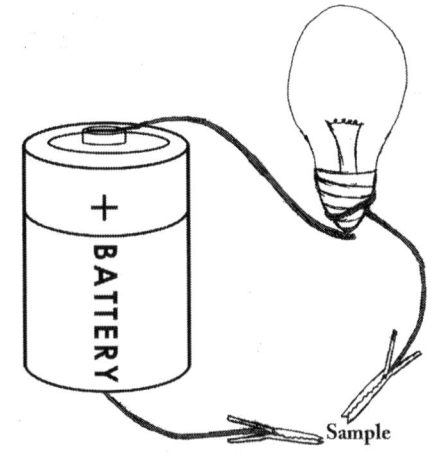

4. Draw conclusions and add them to the science notebook.

Troubleshooting

If you are having trouble getting the wires to wrap around the bulb and stay put, you may want to try making your own socket using a rubber band and two tongue depressors. Directions for this can be found at the following website:

🖳 http://www.instructables.com/id/Flashlight-Bulb-Socket-for-EXPERIMENTS/

Explanation

Organic compounds that are slightly acidic, such as pickles and lemons, are able to conduct electricity. This is because of the ions present in an acidic solution. Other organic materials, like plants, are able to conduct electricity because of the slight magnetic field that exist within them. For a material to be a conductor, the electrical charge needs to move freely through the substance, and the presence of ions allows the charge to do just that. Of course, using pickles to move electricity throughout our homes would be inefficient, considering how often we would have to change them.

Science Notebook

☞ Write down a short description of what you did for the experiment and add the results. After that, add what you have learned from the experiment.

Online Lab - Ohm's Law
Purpose

The purpose of this online lab is to develop the relationship among voltage, current, and resistance by manipulating these variables in electric circuits and examining the effects.

Pre-Reading

✎ Print and read the section of the workbook for the "Ohm's Law" online lab.

Procedure

✓ Do the lab entitled "Ohm's Law," and answer the questions as you work through the online lab.

Science Notebook

☞ Add the completed workbook pages that you printed to the science notebook.

Events in Science
Current Events

🕐 Find a current events article related to the field of physics, and complete the article summary sheet found on p. 187 of the Appendix. Once you are done, add the sheet to the events section of your science notebook.

Hands-on Activity
Optional Hands-on

✂ Complete the electrical charge worksheet practice in the Appendix on p. 202. Add your work and answers to the science notebook.

Week 8 — Unit 4 (Hands-on Course) — 4-Day

Weekly Topic

→ This week will finish a look at electricity.

	Day 1	Day 2	Day 3	Day 4
Textbook and Experiment	☐ Read *CK-12 Physical Science* Section 23.3.	☐ Read *CK-12 Physical Science* Section 23.4.	☐ Do the experiment "Conduction."	
Writing	☐ Add the vocabulary to the glossary section of your science notebook.	☐ Answer the assigned questions in the reading section of your science notebook.	☐ Record what you have done in the lab section of your science notebook.	☐ Complete the Events in Science assignment.

Supplies Needed

✓ **Experiment** – Light bulb, Copper wire, D-battery, Electrical tape, 2 Alligator clips, Organic material, such as a pickle, lemon slice, cheese, bread, or a leaf

Other Notes

Week 8	Unit 4 (Digital Course)		3-Day
Weekly Topic			
→ This week will finish a look at electricity.			
	Day 1	**Day 2**	**Day 3**
Textbook and Experiment	☐ Read *CK-12 Physical Science* Section 23.3.	☐ Read *CK-12 Physical Science* Section 23.4.	☐ Do the online lab "Ohm's Law."
Writing	☐ Add the vocabulary to the glossary section of your science notebook.	☐ Answer the assigned questions in the reading section of your science notebook.	☐ Record what you have done in the lab section of your science notebook. ☐ Complete the Events in Science assignment.

Other Notes

Week 9 Notes - Magnetism

Textbook Assignments
Reading
📖 *CK-12 Physical Science* Sections 24.1, 24.2

Written
After you finish reading, answer questions #1–4 in section 24.1 and questions #1–5 in section 24.2. File your work in the reading section of your science notebook. Then define the following terms in the glossary section of your science notebook:

- ☐ Ferromagnetic material
- ☐ Magnet
- ☐ Magnetic domain
- ☐ Magnetic field
- ☐ Magnetic force
- ☐ Magnetic pole
- ☐ Magnetism
- ☐ Magnetosphere

Experiment - Magnetic Strength
Materials
- ✓ 2 Different types of magnets, such as a horseshoe magnet and a neodymium magnet
- ✓ Paper clips (20 to 30)
- ✓ Paper
- ✓ Cardboard
- ✓ Thick books

Pre-Reading
Magnets come in all shapes and sizes like bars, horseshoes, discs, or rings. Three key metals have the ability to become permanently magnetized: cobalt, iron, and nickel. This property is known as ferromagnetism, and it is the key to making magnets. In today's experiment, you are going to see if all magnets have the same strength or if their attractive force is dependent upon the materials of which they are composed.

Procedure
1. Use the following tests to determine the strength of each of your magnets:
    - **Paper clip test** – Pick up one paper clip with the magnet. Then attach another one to the bottom of the first, creating a chain of paper clips. Continue to add more paper clips until the chain breaks or you cannot add anymore. Count how many paper clips you added, and record that on your science notebook.
    - **Thickness test** – Have a partner hold a piece of paper with a paper clip on it. Use the magnet to move the paper clip around. Next, switch the paper for a piece of cardboard and repeat. Then switch the cardboard for a thick book. Keep adding books until the magnet can no longer attract the paper clip and move it. Measure the thickness of the material that the magnet could attract through and record that on your science notebook.
2. Draw conclusions and add them to the science notebook.

Explanation

The strength of a magnet's attractive forces depends upon its size and its chemical composition. Ceramic magnets, such as the horseshoe or bar magnet, are made by pressing and baking either barium ferrite or strontium ferrite powder into a shape. This shape is then cooled and magnetized to form a ceramic magnet. These magnets are relatively inexpensive and resist corrosion well, but ceramic magnets are weak. Neodymium magnets are made from a metal alloy mixture of neodymium, iron, and boron. These magnets corrode quickly, but they are 10 times stronger than the same-sized ceramic magnet.

Science Notebook

☞ Write down a short description of what you did for the experiment and add a chart with the results. After that, add what you have learned from the experiment.

Online Lab

☞ There is no online lab scheduled for this week.

Events in Science

Current Events

🕒 Find a current events article related to the field of physics, and complete the article summary sheet found on p. 187 of the Appendix. Once you are done, add the sheet to the events section of your science notebook.

Hands-on Activity

Optional Hands-on

✂ Use two bar-shaped magnets and iron filings to see a magnetic field. Do this by sprinkling the iron filings in a line, and then push the north poles of two bar magnets toward the line. (You should see that the iron filings are pulled toward the north pole of each magnet bending back around the magnet.) You can reset and flip one of the magnets so that you are pushing a south pole and a north pole toward each other to note the difference. You can also use the horseshoe and disc magnets from the experiment to see what kind of magnetic field the magnets create in the iron filings.

Week 9	Unit 4 (Hands-on Course)			4-Day
Weekly Topic				
→ This week will look at magnetism.				
	Day 1	Day 2	Day 3	Day 4
Textbook and Experiment	☐ Read *CK-12 Physical Science* Section 24.1.	☐ Read *CK-12 Physical Science* Section 24.2.	☐ Do the experiment "Magnetic Strength."	☐ Do the optional hands-on science activity.
Writing	☐ Add the vocabulary to the glossary section of your science notebook.	☐ Answer the assigned questions in the reading section of your science notebook.	☐ Record what you have done in the lab section of your science notebook.	☐ Complete the Events in Science assignment.

Supplies Needed

- ✓ **Experiment** – 2 Different types of magnets, such as a horseshoe magnet and a neodymium magnet, Paper clips (20 to 30), Paper, Cardboard, Thick books
- ✓ *(Optional)* 2 Bar magnets, Iron filings

Other Notes

Week 9 — Unit 4 (Digital Course) — 3-Day

Weekly Topic

→ This week will look at magnetism.

	Day 1	Day 2	Day 3
Textbook and Experiment	☐ Read *CK-12 Physical Science* Section 24.1.	☐ Read *CK-12 Physical Science* Section 24.2.	☐ *(Optional)* Do the hands-on science activity.
Writing	☐ Add the vocabulary to the glossary section of your science notebook.	☐ Answer the assigned questions in the reading section of your science notebook.	☐ Complete the Events in Science assignment.

Other Notes

Week 10 Notes - Electromagnetism

Textbook Assignments
Reading
📖 *CK-12 Physical Science* Sections 25.1, 25.2, 25.3

Written
After you finish reading, answer questions #1–5 in section 25.1, questions #1–4 in section 25.2, and questions #1–4 in section 25.3. File your work in the reading section of your science notebook. Then define the following terms in the glossary section of your science notebook:

- ☐ Electric generator
- ☐ Electric motor
- ☐ Electric transformer
- ☐ Electromagnet
- ☐ Electromagnetic induction
- ☐ Electromagnetism
- ☐ Faraday's law
- ☐ Solenoid

Unit Test
Take Unit Test #5 – Energy starting on *CK-12 Physical Science*, p. 349.

Experiment - Simple Electromagnet
Materials
- ✓ D-battery
- ✓ Insulated copper wire – about 3 ft (1 m)
- ✓ 2 to 3 in (5 to 8 cm) Nail
- ✓ Electrical tape
- ✓ Iron filings
- ✓ Paper

Pre-Reading
🖎 All magnets have a magnetic field in which the force of the magnetic attraction can be felt. If we place iron filings near a magnet, the filings will line up in a pattern that shows the field produced by the magnet. In today's experiment, you are going to test to see if you can use electricity to yield the same effect.

Procedure
1. Wrap the nail with the insulated wire in tight coils, leave at least a 6 in (16 cm) tail on each end. Using the tape, attach one end of the wire to the positive terminal on the battery and the other end to the negative terminal. (See the diagram above to visually check your set-up.)
2. Sprinkle some of the iron filings on a piece of paper. Take hold of the

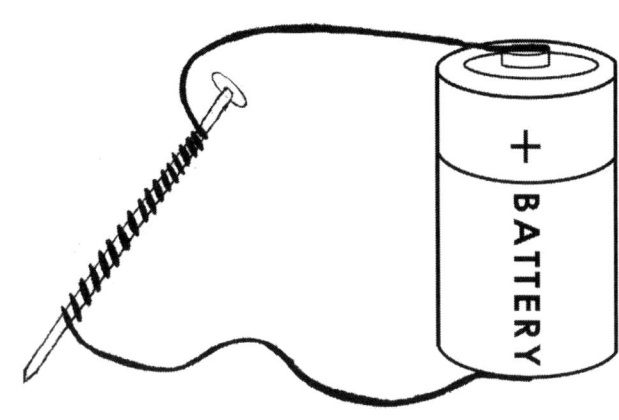

battery in the center and move the wire-wrapped nail close to the filings and observe what happens. If you see a magnetic field form in the iron filings, draw the field on your science notebook.
3. Draw conclusions and add them to the science notebook.

Explanation

Electrical current creates a small magnetic field. When you wrap the wire in a coil around the nail, the magnetic field created is multiplied. The end result is a field similar to that found in a bar magnet, with north and south poles. The magnet created is a temporary magnet because as soon as the electricity (battery) is disconnected, the coil is demagnetized.

Science Notebook

☞ Write down a short description of what you did for the experiment and add the results. After that, add what you have learned from the experiment.

Online Lab

☞ There is no online lab scheduled for this week.

Events in Science

Current Events

🕒 Find a current events article related to the field of physics, and complete the article summary sheet found on p. 187 of the Appendix. Once you are done, add the sheet to the events section of your science notebook.

Hands-on Activity

Optional Hands-on

✂ Make an electromagnetic motor with a battery, 2 safety pins, a rubber band, a bobbin, and copper wire. Watch the following YouTube video for directions:
💻 https://www.youtube.com/watch?v=HHEdIZ282hE

Week 10 — Unit 4 (Hands-on Course) — 4-Day

Weekly Topic

→ This week will look at the periodic table.

	Day 1	Day 2	Day 3	Day 4
Textbook and Experiment	☐ Read *CK-12 Physical Science* Section 25.1 and 25.2.	☐ Read *CK-12 Physical Science* Section 25.3.	☐ Do the experiment "Simple Electromagnet."	☐ Do the optional hands-on science activity. ☐ Take Unit Test #5 – Energy starting on *CK-12 Physical Science*, p. 349.
Writing	☐ Add the vocabulary to the glossary section of your science notebook.	☐ Answer the assigned questions in the reading section of your science notebook.	☐ Record what you have done in the lab section of your science notebook.	☐ Complete the Events in Science assignment.

Supplies Needed

✓ **Experiment** – D-battery, Insulated copper wire – about 3 ft (1 m), 2 to 3 in (5 to 8 cm) Nail, Electrical tape, Iron filings, Paper
✓ *(Optional)* Battery, 2 Safety pins, Rubber band, Bobbin, Copper wire

Other Notes

Week 10 — Unit 4 (Digital Course) — 3-Day

Weekly Topic

→ This week will look at the periodic table.

	Day 1	Day 2	Day 3
Textbook and Experiment	☐ Read *CK-12 Physical Science* Section 25.1 and 25.2.	☐ Read *CK-12 Physical Science* Section 25.3.	☐ *(Optional)* Do the hands-on science activity. ☐ Take Unit Test #5 – Energy starting on *CK-12 Physical Science*, p. 349.
Writing	☐ Add the vocabulary to the glossary section of your science notebook.	☐ Answer the assigned questions in the reading section of your science notebook.	☐ Complete the Events in Science assignment.

Other Notes

Physical Science Unit 4 - Week 10

Unit 4 - Answers

Week 3
Science Notebook Assignment (Experiment)
1. 2.1 m/s
2. 1 m/s
3. 56.72 Hz
4. 0.375 km (or 375 m)

Week 5
Science Notebook Assignment (Experiment)
1. 1.0×10^8 Hz
2. 5×10^7 Hz
3. 15.8 m

Week 8
Science Notebook Assignment (Hands-on Activity)
1. 5400 C
2. 36,000 C
3. 68 amps
4. 13.4 sec

Physical Science

Unit 5 – The Science Fair Project

Unit 5 - Overview of Study

Topics Covered

- **Week 1:** The Science Fair Project, Part 1 – Choosing the topic and Doing Research
- **Week 2:** The Science Fair Project, Part 2 – Formulating a hypothesis and Designing an experiment
- **Week 3:** The Science Fair Project, Part 3 – Performing the experiment and Analyzing the data
- **Week 4:** The Science Fair Project, Part 4 – Wrapping up and Presenting the project

Supplies Needed

Week	Experiment
1	Materials will vary based on the project chosen.
2	
3	
4	

Week 1 Notes - The Science Fair Project
Choosing the Topic and Doing Research

Step 1: Choose the Topic
Explanation
You should choose a topic in the field of science that interests you, such as rockets. Next, come up with several questions you have relating to that topic, (e.g., "How does the design of the rocket affect its flight?" or "What type of rocket flies the highest?"). Then choose the one question you would like to answer and refine it (e.g., "How does the fin design affect the rocket's flight?"). You can get ideas for projects from Janice VanCleave's *A+ Science Fair Projects*.

Key 1: Decide on an area of science.
You should choose an area that fascinates you, something in science that you want to know more about. Begin by brainstorming about things in science that interest you. Rank these areas by degree of interest, and then choose one area to focus on. If your area is too broad, you will want to narrow it down a bit by asking yourself what is interesting about the particular field. For example, let's say you have the following brainstorm list:
- Plants
- Animals
- Light bulbs
- Magnets

Let's say you decide that you are most interested in plants. Plants are quite a broad topic, so to narrow it down a bit, ask yourself:
- What is it about plants that I find interesting?

And you realize that you are most interested in the way that plants grow. So, you write down the following for the area of science you would like to explore with your science fair project:
- How plants grow

You are ready to move onto the next part.

Key 2: Develop several questions about the area of science.
Once you have determined an area, you need to develop several questions about the topic that you can answer with a project. Good questions begin with how, what, when, who, which, why, or where. At this point, you are just thinking of possible questions.

Let's look at some of the questions for the topic, the growth of the plants. For this section, you could come up with questions like:
- Why do plants grow?
- How fast do plants grow?
- When do plants grow?
- Which plants grow faster?

Once you have several options for questions, you are ready to move onto the final section of this step.

Key 3: Choose a question.

Now that you have several options of questions that you can answer with a science fair project, you will need to choose one of those questions for your project. Some of your questions will be easy to develop into an experiment for a science fair project that will determine the answer, but some will not. To help you see this, let's analyze one of the questions our sample student came up with:

> Why do plants grow?

This question is too broad for this assignment because sunlight, water, and nutrients all affect plant growth, not to mention the weather and a whole host of other factors. It will be far too time-consuming to do an experiment that will measure the answer to this question. This student needs to narrow down the question. Some options would be:

> How does the lack of sunlight affect the growth of house plants?
> Which soil is best for house plants to be grown in?

Each of these questions is more specific, making them far easier to measure. It is important to note that you may need to tweak and adjust your topical question as you proceed through the next several steps. You may find through your research that you need to be more specific. Or you may discover as you design your experiment that your question needs to be a little less specific. Your question remains fluid until you begin the experiment.

Step 2: Do Some Research
Explanation

Now that you have a topic and a question for your project, it is time to learn more about your topic so that you can make an educated guess (hypothesis) about the answer to your question. For the question stated above, you would need to research topics like rockets, fin designs, and flight. Begin by looking up the topic in the references you have at home. Then make a trip to the library to search for more on the topic. As you do your research, write any relevant facts you have learned on index cards, and be sure to record the sources you use.

Key 1: Brainstorm for research categories.

This is an important key because developing relevant research categories before you begin to search for information will help you to maintain a more focused approach. It will also help you know where to begin your research and how to determine what information is important to your project and what is not.

You should have at least three categories and no more than five. This will help you to obtain relevant information and make it easier for you to write your report. Let's go back to our sample student with the topical question, "Which soil is best for house plants to be grown in?" You could then come up with the follow topics to research:

> What is found in soil
> Plant growth
> General information about plant structure
> Types of soil

Once you have chosen your research categories, assign each category a number.

Key 2: Research the categories.

Begin by looking at the reference material that you have close at hand, such as the encyclopedias that are available to you in your home. Then you can look to your local library or the internet for additional information.

As you uncover bits of relevant data, write down each fact in your own words on a separate index card. You should number each card at the top left with its category to make the cards easier to organize. We also recommend that you assign a letter for each reference you use, which you can write in the right-hand top corner of each card. This way, after you organize and sort your cards, you will know which references you need to include in your bibliography. Your index cards would look like the one below:

```
┌─────────────────────────────────────────────┐
│ Category Number          Reference Letter   │
│                                             │
│           One piece of Information          │
│                                             │
└─────────────────────────────────────────────┘
```

As you gather a lot of information from your research, it may seem like a waste of index cards, but it is important that you follow this procedure as it will help you organize your research. In the next key, you are going to sort through what you have found and determine what is relevant for your project and what is not. Now, let's go back to our sample student who is interested in how plants grow.

You are researching four different categories and have found a book all about soil, which you label as reference B. You find out that there are several different types of soil, such as sand, silt, clay, loam, peat, and chalk. From another reference, which you label reference G, you discover that a good portion of the world's sandy soil is found in the Great Sahara Desert. So, you would add two cards to your stack that would look like the following cards:

```
┌─────────────────────────────────────────────┐
│ 4                                        B  │
│                                             │
│       There are several different types     │
│       of soil, such as sand, silt, clay,    │
│       loam, peat, and chalk.                │
│                                             │
└─────────────────────────────────────────────┘
```

Physical Science Unit 5 - Week 1

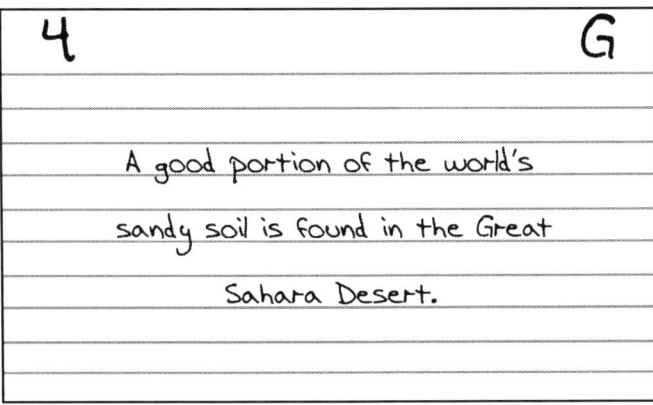

And once you have a stack of index cards, you are ready to move onto next week's assignments.

Week 2 Notes - The Science Fair Project Formulating a Hypothesis and Designing an Experiment

Step 2: Do Some Research
Explanation
This week, you will finish your research, organize your research index cards, and write a brief report on what you have found out.

Key 3: Organize the information.
Once you have finished your research, you need to organize and sort through the information that you have found. Begin by sorting the cards into piles using the research categories in the top left hand of your index card. Then read through each fact and determine five to seven of the most relevant pieces of information from each pile. You will need to decide which facts are relevant to your project (i.e., useful for answering your topical question) and which ones are not. These facts will then form the basis of your report.

Key 4: Write a brief report.
Determine the order of your research categories. Normally, you would go from broad information about the subject to more specific information for the project. After you do this, you need to take the five to seven facts from the first category and turn these into a three to four sentence paragraph by combining the facts into a coherent passage.

Repeat this process until you have a three to five paragraph paper. Then, you need to edit and revise the paper so that it becomes a cohesive report. Finally, you will need to add in a bibliography with the resources you used for your report. Once you have a report written, you can move onto the next step.

Step 3: Formulate a Hypothesis
Explanation
A hypothesis is an educated guess. For this step, you need to review your research and make an educated guess about the answer to your question. A hypothesis for the question asked in step one would be, "The rounder and smaller the fins, the faster the rocket will go."

Key 1: Review the research.
You need to review your research so that it is fresh in your mind. You can do this by reading over your report or by reading over the index cards you made.

Key 2: Formulate an answer.
After you have reviewed your research, you should also read over your question again. Once all the information is fresh in your minds, you are ready to make an educated guess at the answer to your question. Craft a response in the form of an if-then statement that you will be able to design an experiment to test. However, keep in mind that not all questions can be answered easily with if-then statements. As you design your experiment in the next step, you can still make a few adjustments to your hypothesis, if necessary.

So, let's go back to the sample student with the question, "Which soil is best for house plants to be grown in?" Your have done your research and found out that plants depend upon soil for water and nutrients. You believe that the more loam the soil has in it, the more nutrients it will hold, and the better the plant will grow. So, your hypothesis might look like one of the following options:
> If a plant is grown in a potting soil, then it will grow much faster.
> I believe that plants grown in potting soil will grow better than plants grown in other types of soil.

Once you have a hypothesis, you can move onto the next step.

Step 4: Design an Experiment
Explanation
Your experiment will test the answer to your question. It needs to have a control and several test groups. Your control will have nothing changed while your test groups will change only one factor at a time. An experiment to test the hypothesis given above would be to make several different designs fins for your rocket: one that is rounded and small, one that is rounded and normal size, one that is rounded and large, one that is squared off and small, one that squared off and normal size, and one that is squared off and large, plus the standard fins that come with the rocket. If time allows this week, you can go ahead and begin your experiment this week.

Key 1: Choose a test.
Ask yourself what kind of a test you could use that would answer your question and prove your hypothesis either true or false. Write down each idea you have. If you find that you cannot come up with any options for testing your hypothesis, you may need to tweak your statements a bit. If you decide to do this, make sure you verify that the new versions still answer your original topical questions.

Let's go back to our sample student; you could come up with the following ideas for an experiment:
> I could grow a whole bunch of different kinds of plants in sand to see which one would grow the best.
> I could grow one type of plant in several types of soil to see which one would grow the best.
> I could grow several different kinds of plants in several different kinds of soil to see which one would grow the best.

The first option does not really test to see if your hypothesis is true, since it tests the growth of different types of plants in the same soil rather than how the types of soil would affect the growth of those plants. Your third option takes care of the problems in the first, but this experiment would be very involved and would probably take more time than your have. The most logical choice is the second option as it will verify the accuracy of the your hypothesis. Plus, it will also be the easiest experiment test to set-up and perform.

Once you have written down several ideas, review the options and choose one of the ideas for your projects.

Key 2: Determine the variables.

Now that you have chosen a method to test your hypothesis, you need to determine the variables that will exist in your test. There are two main types of variables at play in any experiment: independent variables and dependent variables. Let's take a closer look at each.

- ✓ The independent variable is the factor that is controlled or changed by the scientist performing the experiment.
- ✓ The dependent variable is the factor being tested in the experiment. The dependent variable is what the scientist uses to measure the effect of the changes to the independent variable. In other words, the dependent variable depends upon the independent variable.

It is also important to mention controlled variables, which are factors that are not being examined in the experiment. A scientist will keep the controlled variables constant to reduce their affect on the test.

Once you understand the different types of variables, you need to determine the factors that are at play in your experiment. You can answer the following questions to help you see what the variables are:

- ➢ What factor am I trying to test? (Independent variable)
- ➢ What factor will I use to measure the progress of the test? (Dependent variable)
- ➢ What factors do I need to keep constant so that they will not affect my results? (Controlled variables)

After you have answered these questions, write down your independent, dependent, and controlled variables.

For example, our sample student that will test the hypothesis by growing grass in several different types of soil. Here is what you would write down for the variables in your experiment:

- ➢ Independent Variable — The type of soil
- ➢ Dependent Variable — The growth of the plant
- ➢ Controlled Variables — The amount of sunlight, the amount of water, the size of the pot, the type of grass, and the nutrients in the soil.

Once you have determined the variables for your experiment, you are ready to move onto the next part of this step.

Key 3: Plan the experiment.

Now that you understand the variables at work, you are ready to use this information, along with your testing idea, to create an experiment design. You need to have a control group as well as several test groups. The control group will have nothing changed, whereas each of the test groups will have only one change to the independent variable. You should also plan on having several samples in each of your test groups. Here is what our sample student's test groups could look like:

- ➢ Control Group — a six-inch pot with soil from my backyard
- ➢ Test Group #1 — a six-inch pot with potting soil

> Test Group #2 – a six-inch pot with sand

Next, you choose to use a grass seed blend from the local hardware store as your plant in the experiment, which you note will take up to two weeks to germinate. You decide that you want to measure at least two weeks of growth, so you will need four weeks for your experiment. Now, you can come up with a plan, like this:

> I will begin by filling three pots with soil from my backyard, which I will call my control group. Then I will fill three more pots with potting soil from the store, which I will call test group #1. Finally, I will fill three more pots with sand from the sand box, which I will call test group #2.
>
> Next, I will plant one tablespoon of grass seed in each pot. I will water each of the pots with a nutrient-rich solution made from fertilizer and water on the first day. After that, I will set each of the pots on a window sill in full sunlight. I will check the pots every day for four weeks and water them with the nutrient-rich solution when the soil appears dry. Once I notice that the plants have sprouted, I will record how much they grow each day until the end of the experiment.

Once you have a plan for your experiment, there is just one more thing to do for this step.

Key 4: Review the Hypothesis

Now that you have designed your experiment, you need go back and make sure that your test will prove your hypothesis true or false. Ask yourself, if you got this result from your experiment, would that prove your hypothesis true or false? You may have to ask this type of question several times to verify that your experiment design is sound.

So, for our sample student's science fair project, you would read over your hypothesis and your experiment design. Then you could ask the following question, "If you find that the plants in the potting soil grow five inches taller than the control group, would it prove your hypothesis true or false?" If you say yes, this means that you now have a possible valid experiment design.

Week 3 Notes - The Science Fair Project Performing the Experiment and Analyzing the Data

Step 5: Perform the Experiment

Explanation

This week, you will perform the experiment you designed last week. Be sure to take pictures along the way as well as record your observations and results. (**Note**—*Observations are a record of the things you see happening in your experiment. For instance, an observation would be that the rocket with the wide, rounded fins wobbled quite a bit during flight. Results are specific and measurable. For instance, results would be that the rocket with wide, rounded fins flew 50 feet into the sky and landed 65 feet from the launch stand. Observations are generally recorded in journal form, while results can be compiled into tables, charts, and graphs or relayed in paragraph form.*)

Key 1: Get ready for the experiment.

You already have a plan in place, but there are still a few things you need to do before beginning your experiment. You need to look at a calendar and make sure that you will be home for the duration of the trial because you will need to be there to make observations and record results on each day of testing. You also need to gather and prep any materials that you will be using during you experiment.

So back to our sample student; you will need to block out four weeks where you will be able to check the project daily. You will also need to gather potting soil, sand, soil from yard, nine pots, fertilizer, and grass seed. Finally, you will need to mix up your nutrient rich solution using the directions on the back of the fertilizer you purchased. Once you have everything prepared, you are ready to actually do the experiment.

Key 2: Run the experiment.

You have done a lot of work to reach this point, but that preparation has paved a smooth road for your experiment. At this point, you are familiar with your research and your design, so you should be able to carry out your testing with little to no help. You want to make sure that you write down a list of things you need to check each day during the experiment. Be sure that you include taking pictures of what you see on your list as you will need these images for your project board. For example, our sample student's list could include the following:

> Take pictures of the plants every day.
> Make daily observations about each of the groups of pots in the experiment.
> Once the grass begins to grow, measure and record how much it has grown.
> Check the soil daily for moisture, add water if needed.
> Rotate the pots every week so that both sides receive equal exposure to the sun.

Key 3: Record any observations and results.

As you run your experiment, you need to compile your observations and results. Observations are the record of the things the scientist sees happening in an experiment and results are specific and measurable. Observations are generally recorded in journal form and results can be compiled into tables, charts, or graphs.

Let's go back to the student we have been discussing. You have kept a journal of your observations during the experiment. Your journal contained entries like the two examples below:

> Day 8 - The grass in the pots from test group #1 and the control group has finally sprouted. The pots of test group #2 look like they might sprout tomorrow. I had to water the pots in test group 1 today.

> Day 17 - The grass in the test group #1 pots has grown the tallest so far. This is followed by test group #2 and then the control group. All of the pots have grass that is looking green and healthy. I had to water the pots in test group #1 and #2 today.

You also measured how much the plants had grown each day and then plotted each of your measurements on a graph that looks like this:

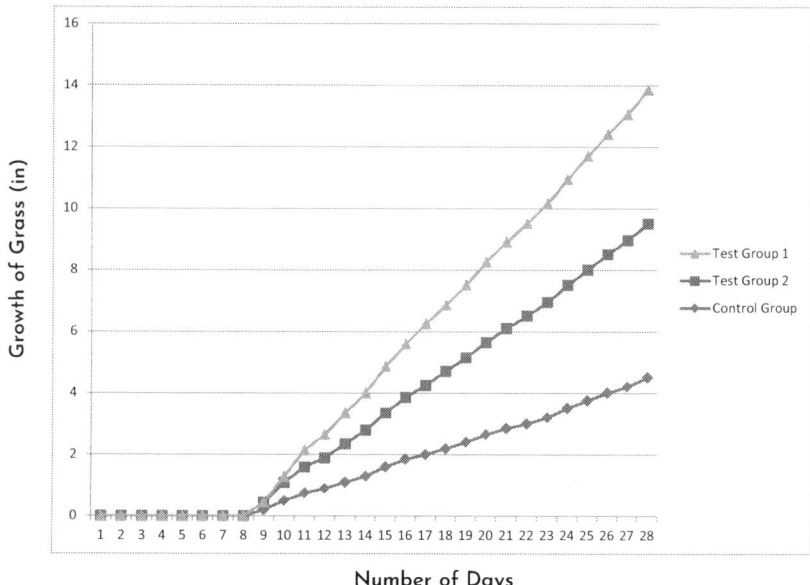

Once you have finished your experiment and have collected your data, you are ready to move onto the next step.

Step 6: Analyze the Data
Explanation

Once you have compiled your observations and results, you can use them to answer your question. You need to look for trends in your data and make conclusions from that. A possible conclusion to the rocket-fin design experiment would be, "I found that the smaller and more rounded the fins, the further and smoother the rocket will fly." If your hypothesis does not match your conclusion, or your were not able to answer your question using the results from your experiment, you may need to go back and do some additional experimentation.

Key 1: Review and organize the data.

Now, you need to analyze your observations and results to determine if your hypothesis are true or false. To do this, you need to read over each of your journal entries and note any trends in your observations. You also need to interpret the charts or graphs you created in the last step and write down the information that you can glean from those.

For the sample project we have been discussing, you note the following trends from your observation journal:

> All of the plants were healthy throughout the test.
> The grass in test group #1 appeared to grow quicker than test group #2 and the control group.
> The pots in test group #2 were water the most frequently.

You also recognizes the following information from your results:

> The grass in test group #1 grew the best.
> The grass is test group #2 grew better than the grass in the control group.

Now, you are ready to state the answer to your science fair project question.

Key 2: State the answer.

Now that you have noted trends from their observations and interpreted information from your results, you can use those data to answer your question. You need to first determine if they have proved your hypothesis true or false. Once your have decided if your hypothesis statements were true or false, you can craft a one sentence answer to your original topical questions from step one.

Your statement should begin with, "I found that ___" or, "I discovered that ___." In the rare case that you are unable to state an answer to your question, you need to take what you have learned, go back to the drawing table, and redesign your experiment.

In our sample project, you are to see from your journal entries, and from the graph you created that your hypothesis was proven true. Remember that the original question was, "Which soil is best for house plants to be grown in?" In this part of the process, you state the following as an answer to your question:

I discovered that the house plants in my experiment grew best in potting soil.

It is important to note that you are theorizing that potting soil is the best medium for growing house plants. This is why you need to begin your statement with, "I discovered" and use the phrase "in my experiment." If you want to learn more about theory versus fact, please read the following article:

🖥 https://elementalblogging.com/theory-vs-fact/

Physical Science Unit 5 - Week 3

Key 3: Draw several conclusions.

When you draw conclusions, you are putting into words what you have learned from your project. Each conclusion should include the following:
- ✓ The answer to your question;
- ✓ Whether or not your hypothesis was proven true (Note: If the hypothesis is proven false, you should state why.);
- ✓ Any problems or difficulties you ran into while performing the experiment;
- ✓ Anything interesting you discovered that you would like to share; and
- ✓ Ways that you would like to expand the experiment in the future.

Our sample student could have written the following conclusion:

I discovered that the house plants in my experiment grew best in potting soil. This finding proved my hypothesis to be true. I also found it interesting that the plants in test group #2 had to be watered more than the other pots. In the future, I would like to test how water affects plant growth or how much moisture different types of soil can hold.

Week 4 Notes - The Science Fair Project Wrapping Up and Presenting the Project

Step 7: Create a Board
Explanation
This week, you will be creating a visual representation of your science fair project that will serve as the centerpiece of your presentation. You will begin by planning the look of your board, move onto preparing the information, and finally pull it all together.

Key 1: Plan out the board.
The science fair project board is the visual representation of your hard work so you definitely want to put as much effort into this step as you have into the others. The board will have specific sections that are set, but you should personalize the look with color and graphics that suit your tastes and match your projects.

The left section of the board typically has:
- ✓ **Introduction** – This includes your question and a brief explanation of why you chose that topic.
- ✓ **Hypothesis** – This is your educated guess from step three.
- ✓ **Research** – This is the report you wrote in step two.

The center section of the board typically has:
- ✓ **Materials** – This is a list of the supplies used in the project.
- ✓ **Procedure** – This is a brief explanation of what was done for the experiment.
- ✓ **Pictures, Graphs, and Charts from the Experiment** –These are visual representations of the project. **Note**—*If you have a highly visual project, you could place the materials and procedure sections on the right flap instead.*

The right section of the board typically has:
- ✓ **Results** – This includes the trends from the observations and the interpretation of the results.
- ✓ **Conclusions** – This is the concluding paragraph from the previous step.

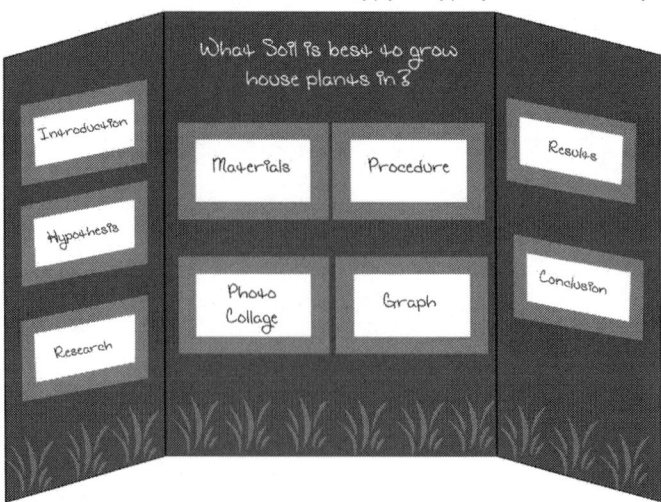

Key 2: Prepare the information.

You have put in a lot of effort until this point, but the work you have done in the previous steps will make it easier for you to prepare the information for your board. You need to type up the information and choose a font type and size for your board.

- ✓ **The Introduction** – Turn your question from step one into a statement. Then write two to three more sentences explaining why you chose your specific topic. You should end your introductory paragraph by sharing the question that you were trying to answer with your project.
- ✓ **The Hypothesis** – Type up and prepare your hypotheses from step three for the project board.
- ✓ **The Research** – Type up and prepare your research report from step two for the project board.
- ✓ **The Materials** – Type up a list of the materials you used for your project.
- ✓ **The Procedure** – Revise the experiment design you wrote in step four so that it is written in the past tense.
- ✓ **The Results** – Turn the trends in the observations you noted and the results you interpreted in step six into a paragraph.
- ✓ **The Conclusion** – Type up and prepare your concluding paragraph from step six for the project board.

Once you have the information all pulled together and planned out, you are ready to make your board.

Key 3: Put the board together.

Now that you have planned out your science fair project boards and prepared the information, you are ready to pull it all together. Cut out the decorative elements, and glue them to the backboards. Then print and cut out your informational paragraphs. For added depth, you can glue the paragraphs onto a foam board before adding the information. Finally, add your titles and the finishing touches to your board.

Step 8: Give a Presentation

Explanation

After you have completed your presentation board, determine if you would like to include part of your experiment in your presentation. Then prepare a five-minute talk about your project. Be sure to include the question you tried to answer, your hypothesis, a brief explanation of your experiment, and the results plus the conclusion to your project. Be sure to arrive on time for your presentation. Set up your project board and any other additional materials. Give your talk and then ask if there are any questions. Answer the questions, and end your time by thanking whoever has come to listen to your presentation.

Key 1: Prepare the presentation.

Once you have finished your project boards, you can begin to work on your presentation. Prepare a brief five-minute talk about your science fair project. This talk should include the question you tried to answer, your hypothesis, a brief explanation of your experiment, the results, and the conclusion to your project. Your outline should highlight the main points that you want to cover for your presentation.

Key 2: Practice the presentation.

Once you have finished preparing your outlines for your talk, practice in front of a mirror. Practice looking at the audience while pointing to the different sections on your project board as you present. Once you feel confident with your presentations, give a practice talk to someone in your family.

Key 3: Share the presentation.

It is important that you present your work to an audience and answer related questions from the group. This will reinforce what you have learned as well as help you to discern how to communicate what you know.

Physical Science
Appendix

Scientist Biography Report Grading Rubric

Spelling (points x 1)

4 points: No spelling mistakes.

3 points: 1–2 spelling mistakes and not distracting to the reader.

2 points: 3–4 spelling mistakes and somewhat distracting.

1 point: 5 spelling mistakes and somewhat distracting.

0 points: > 5 spelling mistakes and no proofreading obvious.

Points Earned _____

Grammar (points x 1)

4 points: No grammatical mistakes.

3 points: 1–2 grammatical mistakes and not distracting to the reader.

2 points: 3–4 grammatical mistakes and somewhat distracting.

1 point: 5 grammatical mistakes and somewhat distracting.

0 points: > 5 grammatical mistakes and no proofreading obvious.

Points Earned _____

Introduction to the Scientist (points x 2)

4 points: Includes thorough summary of the scientist's biographical information and why the student chose the particular scientist.

3 points: Adequate summary of the scientist's biographical information and why the student chose the particular scientist.

2 points: Inaccurate or incomplete summary of one of the scientist's biographical information and why the student chose the particular scientist.

1 point: Inaccurate or incomplete summary of both of the scientist's biographical information and why the student chose the particular scientist.

0 points: No introduction

Points Earned _____

Physical Science - Appendix

Description of the Scientist's Education (points x 2)

4 points: Includes thorough summary of the scientist's education.

3 points: Adequate summary of the scientist's education.

2 points: Inaccurate or incomplete summary of one of the scientist's education.

1 point: Inaccurate or incomplete summary of both of the scientist's education.

0 points: No description of the scientist's education.

Points Earned _____

Description of the Scientist's Major Contributions (points x 2)

4 points: Includes thorough summary of the scientist's major contributions.

3 points: Adequate summary of the scientist's major contributions.

2 points: Inaccurate or incomplete summary of the scientist's major contributions.

1 point: Inaccurate and incomplete summary of the scientist's major contributions.

0 points: No description of the scientist's major contributions and interesting facts of their life.

Points Earned _____

Conclusion (points x 2)

4 points: Explanation of why the student feels one should study the scientist and a summary statement about the scientist.

3 points: Adequate explanation of why the student feels one should study the scientist and a summary statement about the scientist.

2 points: Incomplete or incorrect explanation of why the student feels one should study the scientist and a summary statement about the scientist.

1 point: Conclusion does not have an explanation of why the student feels one should study the scientist and a summary statement about the scientist.

0 points: No conclusion.

Points Earned _____

Final Score = (Total Points/40) x 100%

Total Points Earned _____

Final Score _____

Science in the News

Date: _____

Headline: _____
_____

Authored by: _____

My Summary: _____
_____
_____
_____
_____
_____
_____
_____
_____
_____
_____
_____
_____
_____

My Thoughts: _____
_____
_____
_____
_____
_____
_____

Density Calculations

Density is a property of matter that relates mass and volume. Remember that mass tells you how much matter is in something, whereas volume tells you how much space something occupies. So, in other words, density measures how much matter of a particular substance will occupy a given space. We use the equation below to calculate the density of a solution:

$$\text{Density} = \frac{(\text{mass of water} + \text{mass of salt})}{\text{volume of water}}$$

Sample calculation

So, let's try to calculate the density of a solution that contains ¾ cup of salt in 2 cups of water. We know that 1 tbsp of salt has a mass of 17.1 g and that there are 4 tbsp in ¼ of a cup. We also know that 1 cup of water has an approximate volume of 237 mL and a mass of 237 g. We can use what we know with the formula above to calculate the density of the solution. (**Note**—*You do not need to take into account volume for the salt because it is dissolved in the water, meaning it fits into the spaces within the cup of water, so that the volume change is negligible.*)

$$\text{Density} = \frac{(2 \times 237)\text{g} + (3 \times (4 \times 17.1))\text{g}}{(2 \times 237)\text{mL}}$$

$$\text{Density} = \frac{474\text{g} + 205.2\text{g}}{474\text{mL}} = \frac{679.2\text{g}}{474\text{mL}}$$

$$\text{Density} = 1.43 \, \frac{\text{g}}{\text{mL}}$$

Now, it is your turn to calculate the densities of the solutions you made for the week seven experiment.

A Quick Word about Significant Figures

In chemistry, we use significant figures when we round our answers. Significant figures show us the certainty to which we are confident with an answer. So, if the numbers in my calculations all contain three digits, my answer should contain three digits. There are several rules that govern whether or not a number is significant that you will learn later on in your studies. For now, just remember that all your densities should be rounded to show three digits, just like in the sample calculation.

Density of the yellow solution (no salt)

Density = $\dfrac{+}{}$

Density =

Density of the red solution (2 TBSP of salt)

Density = $\dfrac{+}{}$

Density =

Density of the blue solution (¼ cup of salt)

Density = $\dfrac{+}{}$

Density =

Density of the green solution (6 TBSP of salt)

Density = $\dfrac{+}{}$

Density =

Element Profile Page

Discovered by:
_____

Discovered in:
_____

Common uses: _____
_____
_____
_____

Typically found in:
_____
_____

Interesting Facts:
_____
_____
_____
_____
_____
_____
_____
_____

Periodic Table of the Elements

1	2	3	4	5	6	7	8	9	10	11	12	13	14	15	16	17	18
1 **H** Hydrogen 1.008																	2 **He** Helium 4.003
3 **Li** Lithium 6.941	4 **Be** Beryllium 9.012											5 **B** Boron 10.81	6 **C** Carbon 12.01	7 **N** Nitrogen 14.01	8 **O** Oxygen 16.00	9 **F** Fluorine 19.00	10 **Ne** Neon 20.18
11 **Na** Sodium 22.99	12 **Mg** Magnesium 24.31											13 **Al** Aluminum 26.98	14 **Si** Silicon 28.09	15 **P** Phosphorus 30.97	16 **S** Sulfur 32.07	17 **Cl** Chlorine 35.45	18 **Ar** Argon 39.95
19 **K** Potassium 39.10	20 **Ca** Calcium 40.08	21 **Sc** Scandium 44.96	22 **Ti** Titanium 47.87	23 **V** Vanadium 50.94	24 **Cr** Chromium 52.00	25 **Mn** Manganese 54.94	26 **Fe** Iron 55.85	27 **Co** Cobalt 58.93	28 **Ni** Nickel 58.69	29 **Cu** Copper 63.55	30 **Zn** Zinc 65.39	31 **Ga** Gallium 69.72	32 **Ge** Germanium 72.61	33 **As** Arsenic 74.92	34 **Se** Selenium 78.96	35 **Br** Bromine 79.90	36 **Kr** Krypton 83.80
37 **Rb** Rubidium 85.47	38 **Sr** Strontium 87.62	39 **Y** Yttrium 88.91	40 **Zr** Zirconium 91.22	41 **Nb** Niobium 92.91	42 **Mo** Molybdenum 95.94	43 **Tc** Technetium 98.91	44 **Ru** Ruthenium 101.1	45 **Rh** Rhodium 102.9	46 **Pd** Palladium 106.4	47 **Ag** Silver 107.9	48 **Cd** Cadmium 112.4	49 **In** Indium 114.8	50 **Sn** Tin 118.7	51 **Sb** Antimony 121.8	52 **Te** Tellurium 127.6	53 **I** Iodine 126.9	54 **Xe** Xenon 131.3
55 **Cs** Cesium 132.9	56 **Ba** Barium 137.3	*71 **Lu** Lutetium 175.0	72 **Hf** Hafnium 178.5	73 **Ta** Tantalum 181.0	74 **W** Tungsten 183.9	75 **Re** Rhenium 186.2	76 **Os** Osmium 190.2	77 **Ir** Iridium 192.2	78 **Pt** Platinum 195.1	79 **Au** Gold 197.0	80 **Hg** Mercury 200.6	81 **Tl** Thallium 204.4	82 **Pb** Lead 207.2	83 **Bi** Bismuth 209.0	84 **Po** Polonium [209]	85 **At** Astatine [210]	86 **Rn** Radon [222]
87 **Fr** Francium [223]	88 **Ra** Radium [226]	**103 **Lr** Lawrencium [262]	104 **Rf** Rutherfordium [261]	105 **Db** Dubnium [262]	106 **Sg** Seaborgium [266]	107 **Bh** Bohrium [264]	108 **Hs** Hassium [269]	109 **Mt** Meitnerium [268]	110 **Ds** Darmstadtium [272]	111 **Rg** Roentgenium [272]	112 **Cn** Copernicium [285]	113 **Nh** Nihonium [286]	114 **Fl** Flerovium [289]	115 **Mc** Moscovium [289]	116 **Lv** Livermorium [293]	117 **Ts** Tennessine [294]	118 **Og** Oganesson [294]

*Lanthanides

57 **La** Lanthanum 138.9	58 **Ce** Cerium 140.1	59 **Pr** Praseodymium 140.9	60 **Nd** Neodymium 144.2	61 **Pm** Promethium [145]	62 **Sm** Samarium 150.4	63 **Eu** Europium 152.0	64 **Gd** Gadolinium 157.3	65 **Tb** Terbium 158.9	66 **Dy** Dysprosium 162.5	67 **Ho** Holmium 164.9	68 **Er** Erbium 167.3	69 **Tm** Thulium 168.9	70 **Yb** Ytterbium 173.0

**Actinides

89 **Ac** Actinium [227]	90 **Th** Thorium 232.0	91 **Pa** Protactinium 231.0	92 **U** Uranium 238.0	93 **Np** Neptunium [237]	94 **Pu** Plutonium [244]	95 **Am** Americium [243]	96 **Cm** Curium [247]	97 **Bk** Berkelium [247]	98 **Cf** Californium [251]	99 **Es** Einsteinium [252]	100 **Fm** Fermium [257]	101 **Md** Mendelevium [258]	102 **No** Nobelium [259]

Nomenclature Worksheet

Nomenclature, or naming compounds in chemistry, came about when scientist wanted a consistent way to refer to the compounds they were discovering. There are several basic things we need to know before we can learn how to name compounds in chemistry.

First, we use prefixes to distinguish compounds that have the same two elements in differing numbers. The first six prefixes, which are derived from Latin, are *mono* ("one"), *di* ("two"), *tri* ("three"), *tetra* ("four"), *penta* ("five"), and *hexa* ("six"). The prefix *mono*, is usually left out from the beginning of the first word of the name.

Second, many compounds have common names which we use instead. Over the years, it has become traditionally accepted to use these names over their systematic names. For example:

- C_2H_6 – Ethane instead of dicarbon hexahydride;
- H_2O – Water instead of dihydrogen oxide;
- NH_3 – Ammonia instead of nitrogen trihydride.

Finally, polyatomic ions, or ions with more than one atom, have special names. For example:

- SO_4^{-2} – Sulfate
- OH^- – Hydroxide
- NO_3^- – Nitrate
- CO_3^{-2} – Carbonate
- PO_4^{-3} – Phosphate

When naming compounds, we use a different set of rules to name ionic and covalent compounds. Here are the rules to follow:

Ionic Compounds
(*metal and nonmetal*)

1. Write the name for the cation, which is the metal ion.

2. Write the name for the anion, which is the nonmetal ion or polyatomic ion. (**Note**—*If the anion is a nonmetal atom, use the root name and add the suffix –ide.*)

Examples
- $NaCl$ – Sodium chloride
- HCl – Hydrogen chloride
- $CaSO_4$ – Calcium sulfate

Covalent Compounds
(*two nonmetals*)

1. Write the name of the first element. If the symbol is followed by a subscript of two or more, use the proper prefix to show the number.

2. Write the root name of the second element and add the suffix –ide. If the symbol is followed by a subscript of two or more, use the proper prefix to show the number.

Examples
- CO – Carbon monoxide
- N_2O_4 – Dinitrogen tetroxide
- PCl_3 – Phosphorus trichloride

Physical Science - Appendix

Nomenclature Practice

Ionic Compounds

MgO _____

NaBr _____

Li_2S _____

$MgSO_4$ _____

$Be(OH)_2$ _____

$Sr(NO_3)_2$ _____

Covalent Compounds

CO_2 _____

NO_2 _____

SO_3 _____

N_2S _____

BF_3 _____

P_2Br_4 _____

Chemical Equations Worksheet

How to Balance a Chemical Equation - Part 1

When we balance a chemical equation, we are making sure that the elements involved are present in equal amounts on both sides of the equation. Let's look at the equation below:

$$H_2 \text{ (Hydrogen)} + O_2 \text{ (Oxygen)} \longrightarrow H_2O \text{ (Water)}$$

We have:

Elements	Reactants	Products
Hydrogen (H)	2	2
Oxygen (O)	2	1

As you can see there are 2 hydrogen atoms on the reactants' side of the equation and 2 hydrogen atoms on the products' side of the reaction. However, there are 2 oxygen atoms on the reactants' side of the equation and only 1 oxygen atom on the products' side of the reaction. This makes the equation unbalanced. We can not change the subscripts in the molecule, or it will change the molecule itself so we need to add a coefficient to balance the equation. A coefficient is the big number in front of the element symbol that tells us how many molecules there are in the equation.

$5O_2$ — coefficient (5), subscript (2), elements symbol (O)

So, what if we added a coefficient of 2 before each of the molecules involved? This would give us:

$$2H_2 \text{ (Hydrogen)} + 2O_2 \text{ (Oxygen)} \longrightarrow 2H_2O \text{ (Water)}$$

Now we have:

Elements	Reactants	Products
Hydrogen (H)	4	4
Oxygen (O)	4	2

The equation is still not balanced. What if we added a coefficient of 2 only to the hydrogen and water molecules?

$$2H_2 \text{ (Hydrogen)} + O_2 \text{ (Oxygen)} \longrightarrow 2H_2O \text{ (Water)}$$

Now we have:

Elements	Reactants	Products
Hydrogen (H)	4	4
Oxygen (O)	2	2

Our equation is now balanced.

If you noticed in the equation above, we didn't write a coefficient if we only had one molecule. This is because we don't normally write a coefficient of 1; instead, the absence of a coefficient implies that there is only one of that molecule needed for the reaction. Now it's your turn to try balancing a few chemical equations.

Balancing Chemical Equations - Part 1 Practice

Add the necessary coefficients to balance the following equations. Use the charts to help you determine which coefficients you need to use.

1. ____ AgNO$_3$ + ____ Cu ⟶ ____ Cu(NO$_3$)$_2$ + ____ Ag

Elements	Reactants start	Products start	Reactants final	Products final
Silver (Ag)	1	1		
Nitrogen (N)	1	2		
Oxygen (O)	3	6		
Copper (Cu)	1	1		

2. ____ AlBr$_3$ + ____ K ⟶ ____ KBr + ____ Al

Elements	Reactants start	Products start	Reactants final	Products final
Aluminum (Al)				
Bromine (Br)				
Potassium (K)				

3. ____ LiNO$_3$ + ____ CaBr$_2$ ⟶ ____ Ca(NO$_3$)$_2$ + ____ LiBr

Elements	Reactants start	Products start	Reactants final	Products final
Lithium (Li)				
Nitrogen (N)				
Oxygen (O)				
Calcium (Ca)				
Bromine (Br)				

4. ____ PbBr$_2$ + ____ HCl ⟶ ____ HBr + ____ PbCl$_2$

Elements	Reactants start	Products start	Reactants final	Products final
Lead (Pb)				
Bromine (Br)				
Hydrogen (H)				
Chlorine (Cl)				

5. ____ NaCN + ____ CuCO$_3$ ⟶ ____ Na$_2$CO$_3$ + ____ Cu(CN)$_2$

Elements	Reactants start	Products start	Reactants final	Products final
Sodium (Na)				
Carbon (C)				
Nitrogen (N)				
Copper (Cu)				
Oxygen (O)				

Physical Science - Appendix

Balancing a Chemical Equations for Reversible Reactions - Part 2

Review Practice – Balance the following chemical equation.

1. ____ Al + ____ HCl ⟶ ____ H_2 + ____ $AlCl_3$

Elements	Reactants start	Products start	Reactants final	Products final
Aluminum (Al)				
Hydrogen (H)				
Chlorine (Cl)				

Balancing Chemical Equations for Reversible Reactions

Reversible reactions are also known as equilibrium reactions. The equations for these reactions depict two simultaneous reactions, known as the forward reaction and the reverse reaction. When balancing these equations, you will follow the same procedure as before.

Practice – Add the necessary coefficients to balance the following equations. Use the charts to help you determine which coefficients you need to use. (**Note**—*When completing the charts for balancing these reactions, use the forward reaction, meaning that the left side of the equation contains the reactants and the right side of the equation contains the products.*)

2. ____ N_2 + ____ O_2 ⇌ ____ NO_2

Elements	Reactants start	Products start	Reactants final	Products final
Nitrogen (N)				
Oxygen (O)				

3. ____ NO_2 ⇌ ____ N_2O_4

Elements	Reactants start	Products start	Reactants final	Products final
Nitrogen (N)				
Oxygen (O)				

4. ____ SO_3 ⇌ ____ SO_2 + ____ O_2

Elements	Reactants start	Products start	Reactants final	Products final
Sulfur (S)				
Oxygen (O)				

5. ____ N_2 + ____ H_2 ⇌ ____ NH_3

Elements	Reactants start	Products start	Reactants final	Products final
Nitrogen (N)				
Hydrogen (H)				

Resultant Force Worksheet

Introduction

Objects have more than one force acting on them at any given time. If the forces are in the same direction, they add together. The effect of this addition on the object would be to accelerate, or move, it in that direction. If the forces are in opposing directions, they subtract or cancel each other out. The effect on the object depends upon the size of the opposing forces. If the two forces are equal, they will balance each other out, and the object will remain still. If one of the opposing forces is greater, the end result will be for the object to accelerate, or move, in that direction.

We can determine how an object will move by calculating the resultant force. The resultant force, which is also known as the net force, is the overall force acting on an object after all the forces are combined. To calculate the resultant force, we use vector quantities to represent the forces. These vectors have both direction and size.

Sample Problems

Here are two sample problems for calculating the resultant force using vectors:

Resultant Force = 5N

The object will continue in the same direction.

Resultant Force = -4N

The object will begin moving in the opposite direction.

Problems

1.

Resultant Force = _____

The object will _____

2.

Resultant Force = _____

The object will _____

3.

Resultant Force = _____

The object will _____

4.

Resultant Force = _____

The object will _____

Physical Science - Appendix

Second Law of Motion Worksheet

Introduction
Newton's second law of motion states that:

> A force acting on an object will change its motion. The greater the force on an object, the greater the change in its motion.

This law can also be written as an equation which looks like:

> Force (F) = mass (m) • acceleration (a)

We can use this equation to calculate the acceleration of an object when we know its mass and the force that has acted on it.

Sample Problems
Let's say you are pushing a grocery cart full of oranges that has a mass of 35 kg with a net force of 70 N. What would the acceleration of the cart be?

$F = 70$ N (kg • m/s^2) $\qquad$ $m = 35$ kg $\qquad$ $a = ?$

Substitute and solve.

70 N $= (35$ kg$) a$

$a = \dfrac{70 \text{ N (or kg • m/s}^2)}{35 \text{ kg}}$

$a = 2$ m/s^2

Problems
Now it is your turn to try a few problems!

1. A girl kicks a 2.5 kg soccer ball with a net force of 20 N. What would the acceleration of the ball be?

2. A truck pulls a trailer with mass of 4000 kg up a hill at a rate of 40 m/s^2. What is the net force that acted on the trailer?

3. A 12.5N force accelerates a boy in a toy car at 0.5 m/s^2. What is the mass of the boy?

Power Worksheet

Introduction

An engine's ability to move a vehicle depends upon how powerful it is. Power measures the rate at which the engine does work. We use the following equation to calculate the power needed:

$$P = \frac{W}{t}$$

The unit for power is Watts (J/s). (**Note**—*Often, in a power problem, we are given the force and the distance an object is moved over a given time. Remember that work is equal to the force times the distance.*)

Sample Problems

You exert a force of 67 N to lift a stack of books to a height of 0.7 m in a time of 3 seconds. How much power did you use to lift the books?

$W = F \cdot d = (67)(0.7) = 47 \text{ J}$ $t = 3 \text{ s}$ $P = ?$

Substitute and solve.

$$P = \frac{47}{3}$$

$P = 15.7$ Watts

Problems

Now it is your turn to try a few problems!

1. An athlete rows a boat across a still pond, doing 5000 J of work in 40 seconds. What is the rower's power output?

2. A weightlifter raises a barbell 1 m with a force of 100 N in 1.3 seconds. How much power did he use to lift the weights?

3. A truck pulls a trailer at a constant velocity for 387 m while exerting a force of 520 N for 1.5 minute (90 seconds). Calculate the power.

Physical Science - Appendix

Mechanical Wave Worksheet

Introduction

If you remember from the motion unit, the speed of an object equals distance divided by time. In the same way, we can measure the speed of a mechanical wave by taking its length and the time it takes to complete a period. The period of a wave is the time it takes to go from one crest of the wave to the next crest of the wave.

The period of a wave is very tiny, so we use the wave's frequency instead to calculate the speed of a wave. Frequency measures the number of periods (or vibrations) a wave completes in one second. So, to calculate the speed of a wave, we use the following equation:

$$\text{Speed } (v) = \text{Wavelength } (\lambda) \cdot \text{Frequency } (f)$$

The SI unit of frequency is Hertz (Hz), which stands for one vibration per second.

Sample Problems

Let's say you have a string on a musical instrument, which you have plucked. The plucking produces a wave with the length of 0.28 m and frequency of 4 Hz. What would the speed of the wave be?

$$\text{Speed } (v) = ? \qquad \text{Wavelength } (\lambda) = 0.28 \text{ m} \qquad \text{Frequency } (f) = 4 \text{ Hz}$$

Substitute and solve.

$$v = (0.28 \text{ m})(4 \text{ Hz})$$

$$v = 1.12 \text{ m/s}$$

Problems

Now it is your turn to try a few problems!

1. A rope is vibrated, producing a wavelength of 3m. The frequency of the wave is 0.7 Hz. What is the speed of the wave?

2. What is the speed of a wave in a spring that has a wavelength of 0.2 m and frequency of 5 Hz?

3. A sound wave at sea level has a wavelength of 6 m. What is its frequency? The speed of sound at sea level is 340.29 m/s.

4. What is the wavelength of an earthquake wave that travels at a speed of 3km/s and has a frequency of 8 Hz?

Electromagnetic Wave Worksheet

Introduction

When we looked at mechanical waves, we used the wave speed equation to calculate their speed or frequency. However, electromagnetic waves all travel at the same speed in a vacuum. That speed is the speed of light, which is 3.0×10^8 m/s. Electromagnetic waves do vary in wavelength and frequency, so we can use the wave speed equation to calculate those. As a refresher, here is the wave speed equation:

$$\text{Speed } (v) = \text{Wavelength } (\lambda) \cdot \text{Frequency } (f)$$

The SI unit of frequency is Hertz (Hz), with stand for one vibration per second.

Sample Problems

So let's say you know that a radio station broadcasts an electromagnetic radio wave with a wavelength of 2.6 meters. What is the frequency of the wave?

Speed $(v) = 3.0 \times 10^8$ m/s   Wavelength $(\lambda) = 2.6$ m   Frequency $(f) = ?$

Substitute and solve.

$3.0 \times 10^8 = (2.6) f$

$f = \dfrac{3.0 \times 10^8}{2.6}$

$f = 1.2 \times 10^8$ Hz

Problems

Now it is your turn to try a few problems!

1. A cable station satellite sends an electromagnetic wave with a wavelength of 3 m. What is the frequency of the wave?

2. What is the frequency of an electromagnetic wave that has a wavelength of 6 m?

3. What is the wavelength of an electromagnetic wave that has a frequency of 1.9×10^7 Hz?

Electrical Charge Worksheet

Introduction

When current flows for a set amount of time, we can calculate the charge. This charge is reported in Coulombs (C) – 1 Coulombs (C) is equal to 6.242×10^{18} electrons). The equation to calculate electrical charge is:

$$Q = I \cdot t$$

The Q stands for electrical charge, which is measured in Coulombs. The I stands for current, which is measure in amperes, and t stands for time measured in seconds.

Sample Problems

So, if there is a current of 26 amperes in a circuit for 3 minutes, what quantity of electric charge flows through the circuit?

I = 26 amps   t = (3 min)(60 sec/min) = 180 sec   Q = ?

Substitute and solve.

$$Q = (26)(180)$$
$$Q = 4680 \text{ C}$$

Problems

Now it is your turn to try a few problems!

1. If there is a current of 45 amperes in a circuit for 2 minutes, what quantity of electric charge flows through the circuit?

2. If there is a current of 10 amperes in a circuit for 1 hour, what quantity of electric charge flows through the circuit?

3. How much current must there be in a circuit if 476 coulombs flows past a point in the circuit in 7 seconds?

4. How much time is required for 43 coulombs of charge to flow past a point if the rate of flow is 3.2 amperes?

Physical Science - Appendix

Physical Science - Appendix

Made in the USA
Columbia, SC
30 August 2023